# THE NEXT SUPERCONTINENT

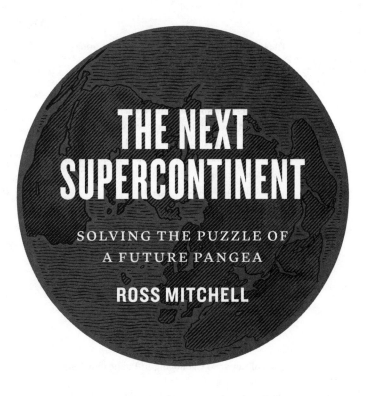

# THE NEXT SUPERCONTINENT

## SOLVING THE PUZZLE OF A FUTURE PANGEA

### ROSS MITCHELL

THE UNIVERSITY OF CHICAGO PRESS

*Chicago and London*

The University of Chicago Press, Chicago 60637
The University of Chicago Press, Ltd., London
© 2023 by The University of Chicago. Illustrations
Copyright © 2023 by Matthew Green

Published 2023
Printed in the United States of America

32 31 30 29 28 27 26 25 24 23     1 2 3 4 5

ISBN-13: 978-0-226-82491-8 (cloth)
ISBN-13: 978-0-226-82492-5 (e-book)
DOI: https://doi.org/10.7208/chicago/9780226824925
.001.0001

Library of Congress Cataloging-in-Publication Data

Names: Mitchell, Ross (Geophysicist), author.
Title: The next supercontinent : solving the puzzle of a future
Pangea / Ross Mitchell.
Description: Chicago : The University of Chicago Press, 2023. |
Includes bibliographical references and index.
Identifiers: LCCN 2022044961 | ISBN 9780226824918 (cloth) |
ISBN 9780226824925 (ebook)
Subjects: LCSH: Plate tectonics. | Geodynamics. | Pangaea
(Supercontinent)
Classification: LCC QE511.5 .M58 2023 | DDC 551.1/36—dc23/
eng20221230
LC record available at https://lccn.loc.gov/2022044961

To Joe and Pete for taking a chance on a Young Turk

# CONTENTS

# PREFACE

Who controls the past controls the future.

GEORGE ORWELL

This is a book about supercontinents—landmasses from which today's continents were born. It explains the evidence for supercontinents' existence and the predictions of a next supercontinent, expected to form some 200 million years from now. This is also a book about how science is done. The scientific method is iterative. The next supercontinent is a scientific hypothesis, and a scientific hypothesis is nothing if it is not tested. As we cannot wait around 200 million years to see if our models turn out to be right, we must use the geologic past to test our ideas about the future.

This will also necessarily be a story about scientists. A mentor of mine insists on including pictures of scientists when explaining scientific concepts as a reminder of the people behind the theories. Some scientists are willing to take their hypotheses to the grave, even in the face of dissent. As the German physicist Max Planck is often paraphrased as saying, "Science advances one funeral at a time." Others are willing to accept new evidence while they're still alive—to change their minds and to put their stamp of approval on competing ideas.

As a scientist myself, I ascribe to the hopefully balanced philosophy that it's okay to like your hypothesis if you love testing it. In high school, I loved all the sciences, particularly chemistry and physics, but wasn't so much a fan of the hours of indoor laboratory experiments. When I arrived at college and heard rumors from classmates that Introduction to Geology blended all the sciences and required fieldwork outdoors, I was immediately sold. When I showed up to the first field trip donning a fresh pair of Carhartt pants—the centerpiece of the geologist's uniform replete with a loop for holding a rock hammer and a back pocket wide enough for a standard waterproof field book—my professor said I was destined to be a geologist. How right she was.

Since that fateful day, I have never wavered from the path of learning more about Earth. I went on to earn my PhD in geology and geophysics at Yale University, did my postdoc at Caltech, was a research fellow in Australia, and am now a professor in Beijing, China, at the Institute of Geology and Geophysics, Chinese Academy of Sciences. Studying the drift of the continents over billions of years, I have conducted fieldwork on every continent except Africa. Rocks scattered around the globe are sample sites targeted for solving the next big problem. Time in the field collecting rock samples makes for time back in the laboratory making careful measurements. But a majority of my time is spent reading, thinking, and writing, and thinking some more (why we call it re-search!). My goal is to integrate plate tectonic theory with an understanding of the whole planet, whereas now it is largely a description of the motions of plates at Earth's surface. To do so, I make an effort to expand my own thinking to address the work done by other geologists that study rocks formed at greater depths of the planet and that are subsequently exhumed to the surface, as well as the work of geophysicists and geochemists investigating Earth's rocky mantle and its molten core.

You may also come away from this book with a better understanding of geology. "Geoliteracy," to borrow and bastardize the term from *National Geographic*, is more important than ever. Although their use of the term focuses on geography, even geography changes as the world changes over time, with once wet climates becoming dry, which is partially why our natural resource needs are shifting toward renewable sources. We need to augment geography—the study of our planet, its atmosphere, and human activity—with geology—the study of our planet's entire physical structure. Climate change is our new reality. I believe one reason for the apparent gap in public and political (if not scientific) opinion on the matter of global warming is the lack of understanding. Geology was only an elective in school when I grew up. How can we have a meaningful discussion about climate when many have so little basic knowledge about how Earth works? Humanity's evolution is now understood to have a close relationship to plate tectonics and climate change. Although the tectonically carved deep lakes of the East African Rift provided a cradle for some of our earliest civilizations, when these lakes dried up due to earlier climate shifts, our ancestors' entire way of life needed to change—they needed to walk distances previously unimaginable to find habitable land.

The polar ice caps that our species has always taken for granted in Antarctica and Greenland are melting at an unprecedented rate. The increasing atmospheric concentration of greenhouse gases begets global warming, which begets ice melting, which begets higher sea levels, warmer oceans, and a climate never experienced by our species. Understanding how today's global climate change will affect our way of life therefore starts with understanding the plate tectonics that underlies why, for example, we have polar ice caps in the first place. Until humans came on the scene, plate tectonics predominantly controlled the concentration of greenhouse gases in the atmosphere, through volcanic emissions. This volcanism is in turn due to the motion of Earth's tectonic plates. Understanding how plate tectonics has caused swings between greenhouse and icehouse climates in the past is therefore critical to understanding what our own emissions might be capable of as well as how we may be able to mitigate warming.

This brings me to my ultimate goal for the book: to bring you up to date on the current state of knowledge of plate tectonics. The large-scale appearance of our Earth's surface will change very little over our lifetimes. But over the billions of years of geologic time, these changes have been immense. I hope this book will provide you with a better recognition of what this geological force has created: from the 9,000-meters high of Mount Everest to the 11,000-meters low of the Marianas Trench. And it hopes to offer a better appreciation of how land masses, now thousands of kilometers away from each other, can merge to form the next supercontinent—when a majority of Earth's continents will come together to form a large, long-lived landmass.

If our species lives to see the next supercontinent, then we will have achieved what no other mammal has ever done: survive more than a few millions of years. Our oldest immediate ancestors, the primate species of the genus *Australopithecus*, lived only about one million years. The longest-living mammalian species have existed only since we mammals filled the niches vacated by the extinction of the dinosaurs 66 million years ago. If current theories about the extinction of the dinosaurs are correct, then our current dominance may well be a cosmic accident. Species larger than bacteria that have survived for hundreds of millions of years are hard to find. Thus the task ahead, our survival as a species, is unprecedented. And while such longevity may seem far-fetched, doesn't it also sound a lot like us? Look at our list of achievements: domesticating fire, cooking, creating languages, inventing the wheel, discovering mathematics, controlling electricity, exploring space.

The survival of our species long enough to see a new supercontinent take shape means we will also need to overcome the human-induced environmental challenges of climate change. The forces of plate tectonics and of human-induced warming of the planet are comparable. Many argue we are now in the Anthropocene: a period in Earth's evolution defined by human intervention. Geologist-turned-climate-scientist Bob Kopp was one of the first to point out that humans have become a geological force to rival plate tectonics.

Human-caused carbon dioxide emissions, for example, now match those from the world's volcanoes. And many of our "geoengineering" solutions for reversing climate change, such as releasing sulfur into the air to cool the planet or "carbon capture" by growing and burying trees, involve, as we will see, humans behaving the way plate tectonics does. Tectonics is the most fundamental control of climate change, and so it is no understatement to say that to solve climate change means nothing less than adopting a mindset of tectonic proportions.

Understanding what the next supercontinent may look like is surely speculative, as it will not happen in our lifetimes or those of our children, grandchildren, great-grandchildren, or their even more distant descendants. Still, I hope this book will encourage you to reflect and speculate on the changing shape of our planet, and on the importance of looking at our planet's evolution at a timescale much longer than that of individual human existence.

# INTRODUCTION

In grade school, many of us learned how the present continents, scattered around the globe, once fit snugly together. Although each continent appears to have its own particular shape, rewind the tape 200 million years and they fit together, like the pieces of a jigsaw puzzle. "Pangea," coined by plate tectonic pioneer Alfred Wegener, means just that: "all Earth," a time in Earth's past when the majority of the continents assembled into a single plate. But Pangea is just the most recent iteration of what's called a supercontinent. At least two others have come and gone over the 4.5 billion years of our planet's existence—and scientists like me believe there will be more in the future. The next supercontinent will likely take another 200 million years to form, but the continents are undeniably on a collision course. According to one computer model, New York City will crash into Lima, Peru. Plate tectonics is certainly powerful enough to stack one city on top of the other, sending the future's equivalent of skyscrapers into the depths of the ocean to be recycled back down into the hot mantle. Although scientists agree that another supercontinent is coming, we have vastly different opinions on how it will take shape.

In this book, I will layout the leading contenders for the geography of the next supercontinent, explore the modern mysteries that still surround plate tectonics, and explain the science behind predicting how continents move. Alas, predicting the next supercontinent is not as simple as understanding today's movements and pressing fast-forward. Tectonic plates move slowly, at about the same speed that our fingernails grow. But GPS is now precise enough to detect this slow motion. And residents of Pompeii, San Francisco, and Fukushima can tell you that the effects of those movements are hard to perceive—until they are devastating. Volcanoes, earthquakes, and tsunamis are evidence of plate tectonics' power. So is geography— just look at the abrupt bend, or kink, in the chain of the Hawaiian Islands (fig. 1). These islands formed a straight chain of semicontinuous volcanic activity for about 30 million years, until a sudden pivot occurred over the course of a few million years or less. The bend is a record of that pivot. Why did this happen? The tectonic plates are all interconnected, so any change in the movement of one plate causes adjustments in them all. Thirty million years ago, Australia broke away from Antarctica and started its current path north across the Pacific Ocean. Whereas the Pacific plate had been moving directly north before the bend, Australia's breakaway in the western Pacific caused the motion of the Pacific plate to deflect toward the northwest after the bend. No plate is moving alone and each plate interacts with its neighbors along their shared boundaries. Plate tectonics is the dance of all plates and the seven major continents (or eight, depending on how you define them) they carry, constituting a global choreography, with dozens of smaller plates in between.

The earliest understanding of plate movement was the sixteenth-century idea of "continental drift"—that the continents migrated like rafts slowly into their current position, floating on an imperceptible layer within the earth. But this theory was largely written off because it was not clear what sort of mysterious substratum the continental rafts would be floating on. By the beginning of the last century, we still knew very little about the interior of the earth. Eventually, as seismology—the study of inner Earth using the vibrations generated from earthquakes—developed and submarines were put to good use

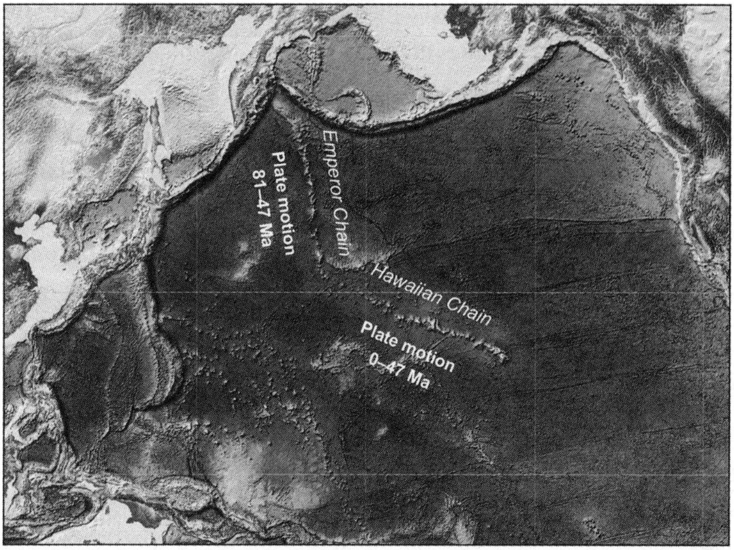

Figure 1. The great kink in the Hawaiian-Emperor seamount chain. The chain of volcanic islands is formed as the plate moves over a stationary hot plume in the underlying mantle. The Pacific plate was moving northward between 81 and 47 million years ago (Ma), then suddenly shifted its drift direction to the northwest, causing the great bend.

after World War II to map the seafloor, the hypothesis of plate tectonics changed geology forever. Breaking Earth's seemingly rigid surface into an interlocking mosaic of different plates that pushed and pulled each other provided a unified explanation for the origin of many of Earth's great geological features, such as mountains, volcanoes, earthquakes, and oceans. But as exciting as the early plate tectonic revolution was, it didn't have all the answers. For example, the hot interior of the earth was likely convecting, in a movement driven by temperature changes, just like the circulation of air in the atmosphere; but how these deep convective cells related to the push and pull of the plates at Earth's surface would remain elusive for decades—and the details of this interaction are still unsolved.

Beneath all the plates, the thickest layer of the planet, the solid but pliable mantle, plays a major role in plate tectonics. Without getting into details of specific hypotheses discussed later, the basic *processes* of plate tectonics according to the current state of knowledge explain plate boundaries, continental drift, opening and closing ocean basins, and planetary cooling through mantle convection. Simply stated, the moving plates are the surface manifestation of mantle convection, and all these processes are linked. Where plates converge, one is thrust beneath the other and sinks back into the interior of the earth. This interaction often occurs where convection in the mantle has a cold downdraft. Where plates are pulling apart, as in the East Africa rift valley today, this is often where convection in the mantle is upwelling as a hot plume.

When on Earth did plate tectonics begin? Have we always had it? Although we take the modern plate tectonic network for granted today, there is mounting evidence that Earth did not always have plate tectonics and that Earth's tectonic style has evolved over time. Indeed, there is fragmentary evidence for plate-tectonic-like processes potentially happening in a very primitive Earth. However, much of this evidence from our most ancient rocks assumes their geochemistry is similar to that of modern rocks and so is interpreted in terms of plate tectonics, but such similarities can also have alternative explanations. If Earth didn't have plate tectonics in its infancy, then how did our young planet behave and how did it evolve in such a way that plate tectonics developed? Putting plate tectonics into the broader context of Earth's long history makes us realize that the current presence of plate tectonics is no guarantee that it will continue forever.

Why? Earth's internal heat budget—the fuel on which our plate tectonic engine runs—is a finite energy source. It is therefore impossible that plate tectonics will operate indefinitely. Maybe a future Earth will look more like the scorched, stagnant lid of tectonically inactive Venus. But rest assured that plate tectonics will last long enough to form another supercontinent, and most likely even a few more cycles after that. But in this book, we won't speculate beyond the next supercontinent just ahead of us. A 200-million-year fore-cast is sufficiently speculative.

While on the issue of prudence in forecasting the future, I should make a note about the scientific approach taken in this book as well as the science that's in it. Predicting the next supercontinent is fertile ground for wild speculation. Scientists have given numer-ous interviews describing their speculations about the nature of the next supercontinent. Indeed, often reasonable rationales are pro-vided to bolster these opinions. But these public speculations hav-en't faced the scrutiny of scientific peer review. With the exception of the book's philosophical last chapter on human survival, we will deal mostly with hard-fought peer-reviewed scientific papers as evi-dence or theory providing the basis for our arguments. The scientific literature will be our check. It's all but inevitable that another super-continent will form, and some obvious signs—the megacontinent of Eurasia is nearly halfway there already—even show that the process is well underway.

No, we cannot test our hypotheses for the next supercontinent by waiting around to see what happens. But because Earth has seen multiple supercontinents in its past, we can use the lessons learned from these previous cycles to test our models for the future. Geology first became a serious science by sticking to its founding adage, a principle called uniformitarianism, that "the present is the key to the past." But now, as practitioners of a scientific field two centuries old, geologists have learned a great deal about the natural experiments that have played out through Earth's 4.5-billion-year history. We now believe that the context of the past is the key to understanding the snapshot of the present—and to project into the future. Geologic his-tory repeats and will continue to do so.

In order for another supercontinent to form, entire oceans must cease to exist. And predicting which oceans will disappear, or "close," and why is a lively debate among geologists. When I was a graduate student, traditional models for the next supercontinent called for closure of either the Pacific or the Atlantic Ocean. Since the Pacific Ocean surrounded (i.e., was external to) the last supercontinent, Pangea, and the Atlantic Ocean is the internal ocean that opened up during the breakup of Pangea, these previous models of the supercontinent cycle were dubbed "extroversion" (for closure of the Pacific) and "introversion" (for closure of the Atlantic), respectively. There are still stalwarts of introversion and extroversion and certainly no consensus yet as to which of these models applies to the next supercontinent. Nonetheless, rejuvenated interest in supercontinent formation is driving new research, and any refined understanding of plate tectonics writ large will be a vast improvement on today's textbooks.

In this book, we will discuss several possibilities for future supercontinents. I will also explain why I have placed my bets on "Amasia," a supercontinent predicted to form at the North Pole. Instead of either the Pacific or the Atlantic Ocean closing, I think evidence points to something more complicated. Close the Caribbean Sea, and the Americas will fuse together; close the Arctic Sea, and the Americas will fuse with Eurasia. The model predicting Amasia is the first to consider the controlling effect of Earth's mantle, the massively thick layer of the planet between the core and the crust, and its massive and forceful convection currents. Even if I fail to convince you that Earth will next host Amasia, this book will leave you with a better understanding of supercontinents and the science behind the very ground beneath our feet.

# I

# PANGEA

The resistance to a new idea increases by the square of its importance.

BERTRAND RUSSELL

The last time the continents were all connected as one landmass, the dinosaurs still roamed the earth. But when supercontinent Pangea *first* took shape about 320 million years ago, even these dinosaurs were still twinkles in their evolutionary ancestors' eyes. It was only shortly before Pangea's formation that animals first climbed onto land from the sea. By the time the continents came together, the first amphibians had spread all over the world. At the same time in the Carboniferous period some 300 million years ago, oxygen rose to its highest levels ever (~30% atmospheric concentration, whereas today it is only ~22%), allowing newly evolved insects to grow so big that dragonflies were the size of watermelons. Elevated oxygen gave rise to the largest animal vertebrates ever, those dinosaurs, who enjoyed their heyday during the peak of Pangea, even after oxygen levels dropped precipitously.

Pangea is the most famous, most recent, and most studied supercontinent. And surely most everything we know about more ancient supercontinents uses what we know about Pangea. And yet we still don't know everything about Pangea. Why it came together and why it broke apart are the most critical pieces of information for answering the question of what the next supercontinent will look like.

We arguably now know with incredible precision *what* continents were included in Pangea, *where* they were positioned, and *when* they assumed this configuration. But *why* Pangea took its particular shape and ultimately *how* it assembled is still unknown.

When I was a graduate student, I read geologist Brendan Murphy and Damian Nance's paper, "The Pangea Conundrum," and was fascinated. Murphy and Nance were identifying a systemic problem surrounding supercontinent research and bringing to light a scientific crisis. I realized they were calling for nothing less than a paradigm shift in supercontinent research. So what was the crisis? Murphy and Nance discovered that the observations didn't match the calculations. Between 600 and 400 million years ago, during the dawn of what's called the "Pangea assembly," there were two ocean basins: the vast and old paleo-Pacific Ocean external to the continents (much like today) and a series of smaller and young oceans (the Iapetus and the Rheic Oceans) that opened at this time and were internal to the continents (like the modern Atlantic Ocean). Theoretical expectations suggested that oceans of old, cold, and dense crust were most likely to vanish by sinking into the mantle. This expectation was further supported by the fact that the young and internal oceans at that time were pushing the continents away from the hot mantle upwelling over which they had opened, causing the continents to drift "downhill" where cold mantle was already downwelling due to convective sinking. Put another way, the continents should have formed together over the paleo-Pacific, and the Iapetus and Rheic Oceans should have expanded. But the accepted shape of Pangea is the opposite—goodbye, Iapetus and Rheic; long live the paleo-Pacific (fig. 2).

Murphy and Nance's conundrum, then, was how to explain why the assembly of Pangea was accomplished by continents consuming their young, hot, buoyant internal oceans only shortly after they

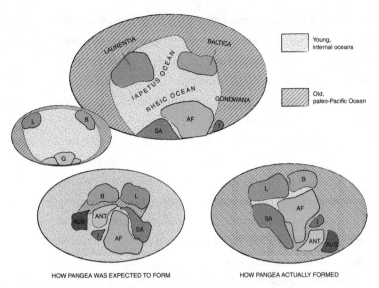

Figure 2. The Pangea conundrum. Top globe shows the positioning of the continents and oceans just before the assembly of Pangea (~420 million years ago), with the old, external paleo-Pacific Ocean and the young, internal Iapetus and Rheic Oceans. On the lower left, Nance and Murphy pointed out that Pangea should have formed by the young internal oceans continuing to spread ever wider at the expense of the external paleo-Pacific Ocean—thus predicting Pangea formation by closing the old ocean. But the opposite situation is what actually happened, which was the reversal of the young oceans from opening to closing—thus Pangea formed by closing the young oceans. AF: Africa; Ant: Antarctica; Aus: Australia; B: Baltica; G: Gondwana; I: India; L: Laurentia; SA: South America.

were created.[1] Why did the motion of the continents quickly reverse itself instead of continuing and consuming the older, colder, denser external ocean? Did something in the mantle change? Which of the assumptions was flawed?

One of the first thing students of plate tectonics learn is to draw a "force diagram" of the system, depicting the pushes and pulls on the plates. Some of the forces come from the plates themselves: "slab pull," where a sinking slab pulls the plate down into the mantle, or "ridge push" where the creation of new ocean crust at the surface pushes the plate away from this spreading ridge. Some of the forces even come from the mantle below the plates. Even though the rocky mantle is stiff because it is under immense pressure (due to lying deeper in the Earth than the crust), it is actually slowly flowing like a very thick syrup because it is so hot close to the core. As a result, the mantle convects, acting like the air in your oven, in which hot air rises as cold air sinks and displaces it. Convective flow in the mantle tends to drag the motion of the overlying plates along, in addition to the pushing and pulling forces acting on the plates themselves.

How these different forces combine determines which direction a plate moves and how quickly. Of the forces within the plates themselves, it is slab pull that is thought to be the strongest. Slab pull is in fact so much stronger than ridge push that the former is often referred to as active and the latter as passive. Ocean crust first forms at an ocean ridge where the hot mantle upwells to the surface. As new mantle upwells and pushes away yesterday's crust, that old ocean crust cools, contracts, and becomes denser. The older and denser the ocean crust, therefore, the more likely it is to sink back into the mantle and vanish, allowing continents to collide. And the forces of the plates can often work in tandem with those of the underlying mantle convection. If you look at where slabs are getting pulled down into the mantle today, it is also where the mantle itself is downwelling due to cold material sinking by convection. In other words, slabs are pulling and the mantle is dragging the continental plates to largely the same places.

But when Murphy and Nance applied these principles to Pangea, they couldn't get it to assemble the way we know it did. The duo were experts on which oceans closed in order to assemble Pangea: the Iapetus and the Rheic, oceans that had once separated North America and Europe from the southern continents of Pangea, known as Gondwana. The only problem was that this separation had occurred only some 550 to 450 million years ago, making the Iapetus and Rheic Oceans likely the youngest, most buoyant oceans around when they vanished shortly thereafter, between 400 and 300 million years ago. To make matters even more confusing, evidence shows that the larger and older paleo-Pacific Ocean, external to the continents, had started to vanish, as expected, but instead of these sinking slabs pulling the continents outward to consume the paleo-Pacific, the continents collapsed inward, consuming their own internal oceans.

As specialized as scientists must become, there is unfortunately quite the divide between those calculating the forces to explain plate motions and those that detect the evidence showing how the motions took place. One of my graduate school advisors, Mark Brandon at Yale University, taught me to regard science as a crossword puzzle. In a crossword, there are words that go vertically ("downs") and those that go horizontally ("acrosses"). For these two types of words to intersect and for the puzzle to be solved, there needs to be an internal consistency between acrosses and downs. By analogy to scientific expertise, there are those who specialize in downs and others that specialize in acrosses. In the case of supercontinents, expert computer modelers who can calculate the myriad forces pushing, pulling, and dragging the plates (geodynamicists) specialize in acrosses. And those that specialize in downs are the rock experts who can tell the story of which continents collided and when (geologists). Essentially, the reason behind this scientific crisis, Murphy and Nance correctly assumed, wasn't that the geodynamicists or the geologists were wrong. It was that they needed to take into account each other's constraints.

Murphy and Nance's contribution was getting these two groups of experts to admit that they were trying to solve the same puzzle of Pangea assembly, and that they needed to work together. Focusing on what the rocks told them, the duo had pointed out a significant problem that needed to be solved. There was something missing in the state of supercontinent research, and nothing less than a paradigm shift was required to overcome it. But before delving any deeper into the problems that still surround Pangea, let us first acknowledge

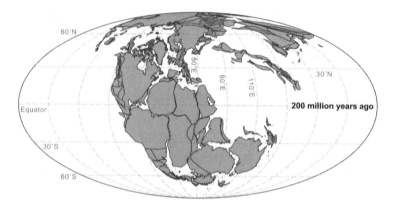

Figure 3. Paleogeographic reconstruction of Pangea 200 million years ago, shortly before breakup.

the scientific crises of the past that we have already successfully overcome as geologists, in hopes that they inspire some momentum to once again look forward to the problems facing us today.

•    •    •

Some geologists might take issue with calling Alfred Wegener the founder of plate tectonics, but none would dispute calling him the founder of Pangea. He was the imagination and deductive mind behind "Pangea" (a term he coined, meaning "all Earth"), which depicted, for the first time, all the world's continents nestled together as a "supercontinent" (translated from *Urkontinent* in Wegener's native German tongue, meaning "first or original continent"). Scientists now believe that several supercontinents have occurred over Earth's history (Pangea, Rodinia, and Columbia), and in my of course somewhat biased opinion as a geophysicist, the supercontinent cycle is one of the grandest theories in our modern understanding of how the earth works. Many details of Wegener's Pangea reconstructions have since been significantly modified, in particular his conception of geologic time periods. But Wegener's geometric configuration of the now famous supercontinent has stood the test of time (fig. 3).

Figure 4. Alfred Wegener, 1930. Photo credit: Johannes Georgi. Photograph courtesy of Archiv für deutsche Polarforschung/Alfred-Wegener-Institut.

Alfred Wegener was a scientist ahead of his time (fig. 4). During the first decades of the twentieth century, Wegener planted the seeds of what would later come to be known as the "plate tectonic revolution" of the 1960s.[2] Although Wegener's concept of continental drift is quite different from plate tectonics, where continents move passively as passengers embedded in plates that in fact move, "drifting" is quite compatible with continental plates flowing "downhill" from what we call "topographic" highs to lows in the underlying mantle. Because Wegener's theory had no viable mechanism, his emphasis on drifting continents contrasted with the fixity of the continents regarded by most in his time. Indeed, even this most basic notion that continents could somehow move was not accepted then by geologists.

His ideas were ahead of the technology available to test them. The year Wegener died, bathymetry measurements of the depth of the seafloor of the Mid-Atlantic Ridge first became available. Bathymetry, the term for topography underwater, is measured using sonar: send sound waves down to the seafloor, count the time it takes them to bounce back to the vessel. You can convert that time to distance by taking into account the speed of sound in seawater. The first bathymetric maps revealed a submerged mountainous ridgeline, the Mid-Atlantic Ridge, that served as a line of divergence away from which Wegener's mobile continents have drifted. Unfortunately, in 1930, on his fourth arctic expedition to measure the thickness of the Greenland ice sheet, Wegener and his team of fourteen men encountered extremely hostile conditions (reaching –60°C) and became stranded without radio contact. Wegener and his companion Rasmus Villumsen pushed ahead, determined to reach base camp to bring food back to the group, but perished from either exhaustion or the conditions.[3] If Wegener had not died at the tragically young age of fifty, he might have identified the evidence for plate boundaries— the mechanics critically missing from his theory of continental drift. If Wegener had seen these ocean ridge data himself, he would have had the evidence to effectively overturn the paradigm of the static continents: he might have completed the revolution he had started.

Still Wegener is one of the most influential figures in the history of geology. He may not have been the first person to notice how the coastlines of South America and Africa were a perfect match, but he was the first person to advance a scientific theory for *why* they fit together. Wegener really only had one contemporary with the same beliefs and foresight, South African geologist Alexander Du Toit. In his book, *Our Wandering Continents* (1937), Du Toit noted that the continents of the southern hemisphere seemed to fit together perfectly, and he went on to muster even more geological evidence than Wegener for the correlation of the southern continents.[4] Comprising South America, Africa, India, Antarctica, and Australia, the southern megacontinent was vast, like modern-day Eurasia. Du Toit dubbed it "Gondwana," deriving the name from a Sanskrit word meaning "forest of Gond," the namesake of a series of shale rocks in India that he had used for geologic correlation to other continents to make his case. The vast Gondwana connection comprises over half of Pangea. But Gondwana wasn't enough for Wegener. He had an entire supercontinent in mind.

•　•　•

Wegener didn't stop with the fit of the coastlines. Wegener went further, offering a means for the continents' movements. If the continents were once positioned together in Pangea, then they must have drifted over geologic time in order to arrive at their present, separated positions. It was the lack of a mechanism that prevented the scientific community from accepting Wegener's theory of continental drift during his lifetime. Without a strong mechanism to explain why continents could be mobile, Wegener had to muster as much empirical evidence for the existence of Pangea and continental drift as he could. And Wegener did just that, drawing lines of evidence from many disciplines within the earth sciences: geodesy, geophysics, geology, paleontology, and of course, climatology, as Wegener was a meteorologist by training. We will explore each of these areas of expertise and why their independent lines of evidence all pointed toward the same conclusion: the continents weren't fixed, but mobile.

Geodesy is the science of measuring the shape and gravity of Earth's surface, and you likely rely on it every day. Think GPS (global positioning system): the device that gives us directions whether we are driving between cities or walking city blocks. Geodesy is also what geologists use to assess the motions on either side of the San Andreas Fault in order to better predict earthquakes. Wegener used geodesy to test whether the continents were drifting very slowly relative to each other. Now of course Wegener did not use GPS, because satellites and rockets to launch them were not yet invented, but the same mathematical triangulation principles applied. Precise geographic locations were determined with ground-based surveying equipment. Indeed, while satellites have increased the precision of our measurements, geodesy from Wegener's day told a similar story: continents move relative to each other at crawling but nonetheless perceptible speeds of a few centimeters per year, about as fast as your fingernails grow. It is impressive, but perhaps not surprising, that Wegener started his case for continental drift with geodesy. Even today, geodetic data are regarded as the most conspicuous and precise evidence that continents move embedded within a rigid network of tectonic plates.

Next Wegener pointed out one of his most fundamental observations: oceans and continents have different elevations. Due to the stark contrast in the densities of their respective compositions, continents (density 2.7 g/cm³) made of buoyant granite ride high, and ocean crust (density 2.9 g/cm³) made of dense basalt is sunken. Like wooden blocks floating in water, these principles of buoyancy apply only if the underlying material, that is, the denser convecting mantle (density 3.35 g/cm³), is soft enough to permit flotation of both continental and oceanic plates. Keep in mind that during Wegener's time, his distinction between continents and oceans, now accepted scientific understanding, was in itself an important discovery. In Wegener's penchant for geophysics as a meteorologist, his distinction between continents and oceans was essentially modern.

When it comes to geologic evidence for Pangea, from the rocks themselves, Wegener describes some of the most salient geologic extremes—the deepest, the tallest, and the most quickly changing geologic landscapes on Earth. He takes us to towering Mount Everest and describes how the Himalayan Mountains formed when India collided with Asia. He takes us to the deep valleys of East Africa and argues that the sinking landscape is the modern expression of the continued rifting of supercontinent Pangea. First, Wegener inferred, North America broke away from Africa, followed by South America, Australia, Antarctica, and India in sequential order, with modern Africa beginning to break apart along the East Africa Rift. In all cases, Wegener correctly choreographed the continental movements involved in the breakup of supercontinent Pangea and the opening of the Atlantic Ocean.

But Wegener's garnering of paleontological evidence from the fossil record is certainly his most successful and lasting legacy. In matching the patterns of fossils on the broken continental fragments of the once contiguous supercontinent Pangea, his insight is entrenched in every first-year geology textbook. The American Museum of Natural History in New York has a cartoon (but accurate!) Pangea puzzle for students to try to solve themselves.[5] And it is essentially the same evidence that Wegener evaluated when he first arranged the continents into the singular snug configuration of Pangea. But this easy-to-digest Pangea puzzle exercise is only half of Wegener's thoughts on biogeography, that is, the geographic distribution of ancient species of animals.

He also used the fossil record to suggest, quite confidently, that the crust underlying the Pacific Ocean had to be older, potentially significantly older, than that of the Atlantic Ocean. We now know that ocean basins, compared to their ancient continental counterparts, are ephemeral geologic features on the surface of the earth, just as Wegener suspected. Ocean basins are overly dense plates that get pushed under, or subducted, when drifting continental plates collide. Now, with the ages of all the world's seafloor dated, the verdict is in. Interestingly, the oldest ages of the crusts underlying the Pacific and Atlantic Oceans are almost identical: about

180 million years old.[6] But that is just the age of the crust. Sure, the Atlantic Ocean is only as old as the age of its oldest ocean crust. But scientists think that a vast external ocean, the ancestor of the modern Pacific basin which is peripheral to all the continents, has been open for hundreds and hundreds of millions of years, even antedating Pangea—again, just as Wegener had suspected based on his biogeography comparisons. Today, we know that such a long-lived "superocean" like the Pacific can be renewed by preferential subduction of its oldest crust and by creation of a new series of oceanic plates within the ocean basin, like a snake shedding its skin throughout its lifetime. The long-term life cycle of such ocean basins and the pattern of the so-called supercontinent cycle are cutting-edge topics of lively debate. Wegener started conversations on countless topics that are ongoing today, and his voice is as relevant as ever.

•   •   •

As prescient as Wegener was, the story of Pangea does not stop there. It would take decades for geologists to test the theory of continental drift—for only by finding conclusive evidence for continental drift could they prove the existence of Pangea. There's a reason that the advances of the 1960s have come to be known as the plate tectonic *revolution*. Even as Wegener insisted on the requirement that the continents somehow drifted, he essentially had no physical mechanism to allow them to do so. Science requires the pillars of both theory and evidence. Despite all the evidence for Pangea and continental drift that Wegener corralled, it wasn't sufficient to be elevated to the status of a scientific theory until it had a plausible physical mechanism to actually make the continents drift. Proving Pangea would await the discovery of plate tectonics.

Although Wegener attempted to explain the origin of both the continents and the oceans, he had data only on the continents and so was most focused on them. But the oceans, it would turn out, hold many of the secrets of plate tectonics. It is under the oceans that most divergent and convergent plate boundaries, the constructive and destructive forces of plate tectonics, are found.

Compared to the ancient continents, the relatively young oceans are transient features on the surface of the earth. Continents are billions of years old, whereas ocean basins are only hundreds of millions of years old. Continents are dominantly comprised of granite, which is relatively buoyant, and ocean crust is dominantly comprised of denser basalt. So when continents collide, it is the ocean crust between them that disappears, plunging beneath continents where it is recycled back into the mantle.

Although oceans may be comparatively short-lived, the information they provide us about plate tectonics is vast. The advent of submarines, and indeed of the Second World War, inadvertently provided the impetus for the plate tectonic revolution. Mapping of the seafloor by submarines for navigational advantages and hiding spots serendipitously gave geologists their first real look at the earth hidden beneath the water. Originally, researchers had expected the seafloor to be flat and featureless. Submarine navigation showed they were wrong, and scientific research institutes continued underwater exploration during the Cold War and during the laying of the first transatlantic telephone cable by Bell Labs of AT&T in the 1950s.[7] The United States boasted three powerhouse oceanographic institutes that were the protagonists of the plate tectonic revolution: Lamont Doherty Geological Observatory of Columbia University (Palisades, New York), Scripps Institution of Oceanography (San Diego, California), and Woods Hole Oceanographic Institute (WHOI; Woods Hole, Massachusetts).

•   •   •

When observing Earth's surface from space, one of the most remarkable features is the occurrence of largely linear mountain chains curving slightly along the edges of each continent. Linear mountain belts are certainly one of the most salient signs of plate tectonics and surely a feature that we would search for on other planets and moons for the possibility of operational plate tectonics in our celestial neighbors.

In fact, the longest mountain chains on Earth are underwater. These submarine ridges are long and interconnected between the world's oceans, wrapping around the planet, much like the seams of a baseball. Some geologists even refer to *an* ocean ridge, since a single underwater ridge can be traced from one side of North America to the other by way of circumnavigating almost the entire globe: the extensive ridge arguably starts in the Arctic Ocean, where it is referred to as the Gakkel Ridge, named for the Soviet explorer who predicted its existence, then continues as the Mid-Atlantic Ridge past Greenland to the south where it bisects the Atlantic Ocean, then connects to the Southwest Indian Ridge which wraps around the Horn of Africa, continuing as the Southeast Indian Ridge around the southern bight of Australia, and connecting to the Pacific Antarctic Ridge and then the East Pacific Ridge and finally terminating on the coast of Baja California. All around the world in six names, but in one continuous submarine ridgeline.

These ocean ridges—these underwater mountain ranges—are the regenerative force of plate tectonics. Ocean ridges occur where the tectonic motions of two plates are diverging, which allows the underlying hot, flowing mantle to reach the surface in a process called seafloor spreading. It was the identification of the towering ocean ridges and their origin in seafloor spreading that kick-started the plate tectonic revolution in the late 1950s and early 1960s. The upwelling of molten mantle, followed by cooling and the formation of new crust, finally provided the means to displace continents away from their original positions snuggled around Africa in Wegener's Pangea and toward their modern locations. Wegner's continental drift finally had an engine.

The heights of mountains on the continents are largely due to their ages; the older the mountains, the lower they generally are, having had more time to erode. Over time, the once towering heights of mountains are razed to the ground by the erosive actions of water, wind, or ice. Bathymetry in the ocean tends to follow the same general rule as topography on land—both deeper and shorter mean older. As molten mantle material cools, it contracts, becoming denser and subsiding. As new magma is constantly rising beneath the ocean ridges, as the seafloor continues to spread, the now older ocean crust gets pushed farther from the heat source, and it cools and contracts as it does so, becoming even denser. In addition to increasing density due to cooling and contracting, ocean crust also becomes hydrated over time, incorporating more seawater into its pore spaces. These processes combine to increase the density of ocean crust as it ages, causing it to move deeper and deeper as it gets older. Eventually it becomes dense enough to sink back into the mantle from which it came. It is nature's great recycling project.

All ocean ridges are approximately the same height, 8,530 feet (2,600 m) below sea level, a height that is set by the temperature of the magma being released from the mantle into the ocean. Although these ocean ridges tend to lose their height over time, the slope of the seafloor away from the ridge changes markedly from ocean to ocean. How is it ridges of the same height could have different slopes? The slope of the seafloor away from an ocean ridge turns out to be a function of the rate of seafloor spreading. And the rate of seafloor spreading away from an ocean ridge is related to the underlying heat source, the mantle. Seafloor slopes in the Pacific Ocean are broad and gentle, and those in the southwest Indian Ocean are narrow and steep, while seafloor slopes in the Atlantic Ocean are between these extremes. The variation in the steepness of seafloor slopes is thought to reflect the different rates at which heat escapes from the mantle in these different oceans. The more heat that is supplied from the underlying mantle, the more magma that is created, the faster the

Figure 5. Marie Tharp mapping the Atlantic seafloor and profiles of the Mid-Atlantic Ridge in 1961. Photograph courtesy of the Lamont-Doherty Earth Observatory and the estate of Marie Tharp.

seafloor spreads, the gentler the seafloor slope. It is therefore not surprising that ocean ridges were first discovered, as we will see next, in the Atlantic Ocean where the slopes away from ridges are less gentle than those in the Pacific Ocean. Put another way, the more active, gentler ridges of the Pacific Ocean were harder to detect.

As early as the late 1940s, geologist, oceanographer, and cartographer Marie Tharp, of Lamont-Doherty Earth Observatory, made some of the first seafloor maps depicting ocean ridges (fig. 5). At the time, misogyny in the form of Navy regulations meant that women weren't allowed aboard ships. Since Tharp wasn't allowed to help gather data until the 1960s, her role was to process the bathymetry data sent back to her at the lab in Palisades, New York. But Tharp went beyond simply processing the data. She helped her colleagues realize the full potential of what the maps revealed: what would come to be known as the Mid-Atlantic Ridge.

Ocean ridges are not known merely for their towering heights, rising more than 6,560 feet (2,000 m) above the base of the ocean floor, about 8,530 feet (2,600 m) below sea level. These hidden underwater ridges are very broad, arching all the way down to the

Figure 6. Diagram of the jagged peaks of the Mid-Atlantic Ridge in the North Atlantic by Marie Tharp and her colleague Bruce Heezen. Reprinted courtesy of copyright © holder Fiona Schiano-Yacopino for Marie Tharp Map of PDNA.

abyssal plains of the seafloor about 10,000–20,000 feet (3,000–6,000 m) below sea level. Tharp's detailed maps indicated that these ridges (fig. 6) are not only elevated, they are also characterized by narrow valleys along their crests. But why? A clue came from one of Tharp's colleagues who found that earthquakes occurred more often near these narrow valleys. Despite the skepticism of her male superiors, Tharp reasoned these seismically active valleys at the top of the ridges could very well be rift valleys formed as the seafloor got pulled apart, thereby allowing the continents to drift apart from each other. Tharp was listed as an author on many early papers of the plate tectonic revolution, but she deserved more credit for her critical role as a mapmaker of the underworld without which the plate tectonic revolution might have been delayed.[8]

Seafloor spreading is not the only way that the overheated mantle cools, and ocean ridges were not the only underwater geologic features discovered by submarines. For the next discoveries of the plate tectonic revolution, we must delve both deeper and shallower. Mount Everest, at 29,029 feet (8,848 m) tall, is the highest elevation on land. The lesser-known lowest point on dry land, at 1,419 feet (433 m) below sea level, is along the shoreline of the Dead Sea between Jordan and Palestine. The difference between the continental high and low therefore is 30,448 feet (9,281 m). Underwater, however, the deepest ocean trench, the Marianas Trench off the coast of the Philippines, was measured by mapping aboard the RV *Kilo Moana* in 2009 at its deepest point, known as Challenger Deep, at 36,070 feet (10,994 m) below sea level. The difference between peaks and valleys underwater is thus 5,622 feet (1,713 m) greater than the difference between them on land. Simply put, bathymetry is more extreme than topography. In a sense, the seafloor is more mountainous than the mountains.

•   •   •

Why are ocean trenches so extremely deep? Ocean trenches are ocean crust pulled downward due to a phenomenon called subduction, the destructive force of plate tectonics. If seafloor spreading occurring at ocean ridges creates new tracts of ocean crust, then it follows that crust must be destroyed in equal measure somewhere else, otherwise the planet would be expanding. Before the plate tectonic revolution took hold, the "expanding Earth" hypothesis—that the volume of our planet was increasing—was in fact entertained as a potential explanation for its crinkled crust (fig. 7).[9] Eventually, a lack of evidence allowed mainstream opinion to dismiss this hypothesis. We've also known since the nineteenth-century work of William Thomson, Lord Kelvin, that the earth is cooling.[10] Since the discoveries of both convection and radioactivity, we now know the planet is slowly losing internal heat sourced both from our fiery planetary formation and from the decay of radioactive elements to icy outer space. With the rare exception of water turning to ice and expanding, most materials contract when they cool. Both theoretically and empirically, therefore, Earth is not expanding. If the constructive

Figure 7. Expanding Earth. An early, now discarded hypothesis to explain the distribution of the fragmented continents (*right*) of the once contiguous supercontinent Pangea (*left*) by the radius of the planet expanding over time. Image courtesy of the public domain from O. C. Hilgenberg, "Vom wachsenden Erdball," *Charlottenburg*, copyright 1933.

force of seafloor spreading is occurring, then an equally destructive force must also somewhere occur. These regions are subduction zones, and their discovery was another cornerstone of the plate tectonic revolution.

Subduction on Earth mainly occurs along what is known as the "ring of fire," a massive rim around most of the Pacific Ocean where a majority of earthquakes, tsunamis, and volcanic eruptions occur (fig. 8). The unfortunate devastation of Fukushima in Japan in March 2011 is recent evidence of the immense risk of living along the ring of fire. The 1964 "Great Alaskan earthquake," also known as the "Good Friday earthquake," is the most powerful earthquake documented in North American history and occurred astride a subduction zone near the southern shore of Alaska on the ring of fire. Earthquakes due to tectonic motion along subduction zones are potentially lethal because they can trigger tsunamis and landslides due to their vertical motion. As plates converge, one is pushed downward and the other is thrust upward. The entire column of water sitting above the uplifted plate is also thrust up, triggering a massively wide swell

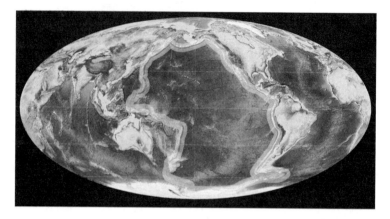

Figure 8. The Pacific Ocean "ring of fire," where most earthquakes, tsunamis, and volcanic eruptions occur today, such as in (going counterclockwise) Chile, California, Alaska, Japan, Indonesia, and New Zealand.

that generates a tsunami wave. This wave loses very little energy as it traverses oceans, and its devastating energy is unleashed on land when the shallowing shoreline makes it crash.

Earthquakes caused by seafloor spreading are shallow (less than 30 km); although many earthquakes due to subduction are also shallow, they can also occur at much greater depth. Since oceans can be vast before they are subducted, when ocean crust is subducted beneath the overriding plate at an ocean trench, the subducted slab of ocean crust hanging down into the mantle may actually reach as far down as all the way to the boundary with the core, traversing the entire massively thick shell that is the mantle. But earthquakes can occur only in brittle rock, so the pattern of deep earthquakes tells us that the subducting slab remains brittle to a depth of nearly 400 miles (600 km), a vivid testament to how anomalously cold it is. Subduction-related earthquakes can therefore be the deepest in the planet. The deeper the earthquake, the wider the potential area that can be harmed because the seismic wave energy propagates farther before dissipating at the surface. But shallow earthquakes due to subduction are actually more lethal, as very little of the seismic energy is dissipated and the population centers affected are simply closer to the earthquake's epicenter. When subduction-related earthquakes are shallow, they affect a smaller area, but the effect is more devastating because the energy is more concentrated: remember the entire towns in Italy destroyed in recent years. Also, the vertical motion of the ocean floor may generate a tsunami.

Earthquakes, tsunamis, and landslides are not the only reasons that living along the ring of fire is particularly perilous. The volcanoes are more volatile there too. As the descending slab of ocean crust is subducted, it dehydrates, releasing the water it absorbed into its pore spaces over millions of years as it migrated away from the ocean ridge. This water is released from the slab and invades the overlying mantle. The water destabilizes the minerals in the mantle, causing some of them to begin melting. The water-rich magma formed initiates a chain reaction of voluminous melting that culminates in the towering volcanic ranges like the Cascades of the northwestern United States. Like a can of your favorite carbonated beverage, the gas-rich magma builds up pressure and explodes as it exploits any weakness it can find in the overlying crust, producing a devastatingly powerful and deadly eruption.

Subduction is the main culprit to trigger earthquakes along the ring of fire, but it is not the only tectonic force at play. The final type of plate boundary to be discovered during the plate tectonic revolution was the "transform fault," which allows plates to slide past each other.[11] Earthquakes along the San Andreas Fault, to pick a world-famous example, prove how transform faults can be as devastating as subduction zones. However, they rarely result in tsunamis because their motion is horizontal and so the energy is not transmitted upward into the seawater—a notable example of artistic license taken by Hollywood in the movie San Andreas as Dwayne "The Rock" Johnson races under the Golden Gate Bridge in a speedboat to crest a "tsunami."

Why is the ring of fire more dangerous than other plate margins? Even though it is the powerful earthquakes of subduction zones and transform faults that are the cause of devastation, the reason for their particular power along the ring of fire may actually have more to do with seafloor spreading. Remember that the ocean ridges of the Pacific Ocean allow more heat to escape from the mantle than anywhere else on Earth. Because of this accelerated rate of heat escape, the ocean ridges of the Pacific Ocean exhibit the fastest rates of seafloor spreading. That is, the Pacific Ocean creates more seafloor than any other ocean. Therefore, even though the ring of fire at the edge of the Pacific is doing its best to consume as much crust as possible due to subduction, the Pacific Ocean is still large because its ocean ridges are rapidly creating new crust. The reason why heat escapes so efficiently from the Pacific Ocean and leads to such prolific seafloor spreading is not well understood, but it most likely relates, as we will discover, to the vanished supercontinent Pangea. Understanding the lethality of the ring of fire, and why plate tectonic boundaries are the way they are, requires an understanding of why and how supercontinents form.

·   ·   ·

Equipped with an understanding of the theory of plate tectonics, we now return to Pangea. Even though Wegener understood that the very existence of Pangea relied on a phenomenon like continental drift, he knew his conception of the displacement forces behind the moving continents was his weakest link. Owning his own emphasis on evidence over theory, Wegener himself admitted, "the [Isaac] Newton of drift theory has not yet appeared."[12] But for theoreticians to take the possibility of continental drift seriously, they had to first be impressed by the body of evidence for Pangea.

This was found in "paleomagnetism," which happens to be my own expertise in geology. The fact that Earth has a magnetic field is one of the planet's unique and redeeming qualities—among other things, that magnetic field shields us from harmful solar and cosmic radiation. And it allows us to measure how the continental positions have shifted over time.

Paleomagnetism is the field of geology that takes advantage of Earth's magnetic field and measures the "fossil" magnetism preserved in some iron-rich rocks when they form. This field was first developed early in the plate tectonic revolution. Submarines during World War II and the Cold War had not only been mapping the bathymetry of the depths of the seafloor looking for hiding spots. They were also towing along devices called magnetometers which would yield huge tactical advantages. Magnetometers had originally been used in aircraft to detect other aircraft, being big hunks of metal with large magnetic signals. Submarines, being even bigger hunks of metal, yield even bigger magnetic signals that could easily be detected. Starting in World War II then, magnetometers were put in submarines to detect other submarines.

During the Cold War, the surplus of magnetometers was used for another purpose: to measure the magnetic response of rock underlying the seafloor. It turned out that there were systematic clues to be found. Magnetic mapping of the seafloor revealed a striped pattern reflecting systematically alternating high and low magnetic field intensity. This striped pattern occurred in essentially straight lines for hundreds or even thousands of miles. (Some scientists now believe whales navigate vast oceans using these magnetic stripes for direction, tracking courses with segments that are either parallel or perpendicular to them.) But the origin of these stripes wouldn't be clear until researchers compared results from a paleomagnetic study of continental rocks.

At the same time that the magnetic stripes of the seafloor were being investigated, paleomagnetists were discovering that continental rocks preserved both normal and reverse magnetic polarities. Eventually these two research worlds collided and resulted in the discovery of seafloor spreading. How did this breakthrough happen? Earth's magnetic field is referred to as a "dipole," with a North Pole and a South Pole. At present, these field lines emerge out of the earth near the South Pole from the core from which they're generated, bend around the equator, and dive down toward Earth's core close to the North Pole. This pattern is known as normal polarity. But there were times in Earth's past when this pattern was the reverse, with field lines emerging out of the earth near the North Pole and diving toward Earth's core close to the South Pole. However, at the time the plate tectonic revolution was about to begin, we did not know that the magnetic field was capable of reversal. First geologists figured out that rocks had minerals in them containing magnetic "domains" strong enough to be measured with magnetometers. Almost immediately, they realized that some rocks were of "normal polarity" (meaning their magnetization aligned with the direction we expect from Earth's present magnetic field), but that an equal proportion also appeared to be of "reverse polarity" (meaning their magnetic fields pointed in the opposite direction of the present magnetic field).

This polarity riddle was the key to solving the magnetization stripes puzzle. When new molten mantle material upwells at seafloor spreading ridges, any magnetic minerals will align their magnetic domains with Earth's magnetic field as the elevated temperature allows them to wiggle around. Then, when the mushy magma eventually cools to form hard rock, the magnetic alignment of the magnetic domains in the iron-rich minerals such as magnetite and hematite gets frozen in time and locked in place. Because seafloor spreading continuously produces new magma along the ocean ridge, the basalts produced passively record the reversals in the Earth's magnetic field as stripes of opposite magnetic polarity. Today, for example, new ocean crust acquires magnetizations that are parallel to Earth's normal magnetic field. Go back about 1 million years, however, a time when Earth's magnetic field was reversed, and all basalt produced at that time was magnetized in the opposite direction, having formed during a period of reverse polarity.

Because the ocean crust was forming on both sides of the active ocean ridge at the same time, these patterns were parallel to and are symmetric about the ridge crest. The fact that the patterns of normal and reversed polarity zones are mirror images on either side of an ocean ridge lent overwhelming support to the notion of seafloor spreading. The first detection of magnetic anomalies of the seafloor was conducted by the Scripps Institution of Oceanography in San Diego, so the first ocean ridge to be imaged in magnetic detail was off the coast of Vancouver Island, now known as the Juan de Fuca Ridge.[13] It was the long-sought smoking gun—the solution to understanding magnetization patterns of the seafloor—and the timing of normal and reverse polarity deduced from this striped pattern matched the timing deduced from the continental record. But it was the magnetizations of rocks on the continents that would directly prove the existence of Wegener's Pangea and his hypothesis of continental drift.

•    •    •

Maybe I chose a less controversial topic for my dissertation work, or maybe science has gotten better at entertaining controversial hypotheses, but the story of how Ted Irving, a pioneer of paleomagnetism, was denied his PhD at Cambridge University is nothing short of shocking by modern standards.[14] In the early 1950s, graduate student Irving made the decision to use the fledgling field of paleomagnetism to test Wegener's Pangea and the controversial continental drift hypothesis. Irving's evidence in support of continental drift was a bridge too far for some of his academic supervisors. Whether their disapproval was rooted in their suspicions about the nascent field of paleomagnetism, or they were simply ignorant of the gravity of his achievements, or they were personally still too resistant to accept such a paradigm shift is difficult to know. Nonetheless, being denied a PhD did not stop Irving.

At Cambridge, Irving had used a newly designed magnetometer in order to measure the latitudes of ancient rocks, following through on a critical test of Wegener's drift hypothesis. Not only do rocks tell us whether the magnetic field was of normal or reversed polarity at the time the rock formed, they can also tell us the latitude at which the continent was positioned at that time. Earth's magnetic field lines are inclined vertically at the poles where they dive in and out of the earth, and between these poles they bend around the planet at different angles, flattening to horizontal at the equator. Today, a lava cooling in Australia will acquire a shallow magnetic "inclination," reflecting the continent's position in the subtropics. But rewind back to a time before Australia broke away from Antarctica, during the heyday of Pangea, and you will find much steeper magnetic inclinations in older lavas, revealing its near-polar location at that time. And this is exactly what Irving discovered.

For his exceptional but underappreciated dissertation, Irving studied the rocks of India: the most obvious test that Wegener had proposed of his continental drift hypothesis. Wegener had long suspected that the towering heights of Mount Everest in the Himalayas were due to India colliding with Eurasia. Like Alexander Du Toit, who shored up evidence for the existence of Gondwana, Wegener saw convincing rock correlations with the southern continents of Africa, Australia, and Antarctica to posit that India, although currently positioned in the northern hemisphere, had originated in the southern hemisphere. The new innovation of paleomagnetism could be used to potentially track this profound northward migration.

Irving had already made enough measurements of local rocks in Scotland to be convinced of the utility of the newly invented magnetometer and of the potential for garnering profound support for the continental drift hypothesis with paleomagnetism. The measured latitudes implied by 1-billion-year-old sandstones in Scotland were much closer to the warm equator than the country's current and relatively frigid latitudes allowed. Irving was convinced he could argue that the inclination data required Scotland to have drifted. He was ready to apply his techniques to a more recent and obvious test of drift: India. Indeed, Irving's dissertation work on rocks in India showed significant changes in magnetic inclination just as India was hypothesized to make its northward migration. As suspected, India moved north before the Himalayan Mountains formed. Even without a PhD in hand, Irving was hired in 1954 by the prestigious Australian National University in order to conduct a similar test of Australia's hypothesized northward migration since breaking away from Antarctica.[15] Ten years later and with more than thirty scientific papers published, in 1965 Irving was finally granted the Cambridge PhD he had deserved all along.

But some of Irving's findings went beyond simply proving Wegener's Pangea and significantly changed our conception of the supercontinent. One weakness of Wegener's reconstructions was that the knowledge of the age of rocks was poor: radiometric dating—backcalculating the age of a rock using the long-term decay of radioactive elements over time—had not yet been invented. Irving's paleomagnetic data proved the existence of Pangea, but their details also called into question how long Pangea had actually existed. Paleomagnetic reconstructions published by Irving in the journal *Nature* painted a much more dynamic picture than Wegener had originally envisioned.[16]

Initial interpretations of the first paleomagnetic data were argued either to invalidate Wegener's Pangea or to suggest the temporary existence of bizarre magnetic fields. Irving proposed a third option that neither invalidated Pangea nor required unusual magnetic fields. Irving's third way was that Pangea had existed, but that it had changed its shape over time—more than Wegener had thought. Irving therefore proposed two Pangeas: Pangea A and Pangea B, where continental drift (by then updated to plate tectonics) had

## Pangea A          Pangea B

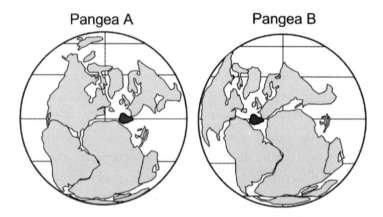

Figure 9. A tale of two Pangeas: Pangea A and Pangea B. Ted Irving proposed that the supercontinent shape shifted from its initial B configuration into the later and final A configuration. Adapted from A. B. Weil, R. van der Voo, and B. A. van der Pluijm, "Oroclinal bending and evidence against the Pangea megashear: The Cantabria-Asturias arc (northern Spain)," *Geology* 29, no. 11 (2001): 991–94.

allowed for the shape-shifting from the B to the A configuration (fig. 9). If plate tectonics had assembled the supercontinent, surely it could also reorganize it.

The Pangea A versus B problem still persists to this day, with some scientists arguing the evidence in support of temporarily unusual nondipole magnetic fields and other hypothesized paleomagnetic aberrations. Still, Pangea's existence is accepted. As Irving conceived, scientists agree the discrepancies are most likely due to minor tectonic motions close to the resolution of the paleomagnetic method itself. The magnetic North Pole is not the geographic North Pole. This is why every navigational chart or map has a magnetic declination indicated in its margins, which is the number of degrees, east or west, that magnetic north is deflected away from pointing to geographic north. Also, magnetic north moves around over time. This is why you need updated maps. Because magnetic north is not necessarily true north, furthermore, magnetic declination is different from place to place.

Luckily, magnetic north averages out to coincide with geographic north over enough time. Because magnetic north wobbles around as much as 30° away from true north over a time period of a few thousands of years, paleomagnetism must sample rocks over this duration and then calculate the *average* latitude. Only when an average is taken does magnetic north match true north. Every measurement scientists make has inherent uncertainty. The uncertainty of the measured average magnetic pole, about 10° (~1,100 km), is about the size of the debated difference between Pangea A and B, as well as the magnitude of other hypotheses including atypical magnetic fields to explain the minor discrepancy. Because of the uncertainty, then, we cannot distinguish between these hypotheses. I should clarify that this debate between Pangea A vs. B is different from the Pangea conundrum that Brendan Murphy and Damian Nance pointed out— but they are related. Sometimes, resolving more tractable problems like Pangea's configuration can provide clues for the more challenging problems such as the geodynamic explanation for why Pangea assembled.

So someday, and hopefully someday soon, we will have resolved the Pangea A vs. B controversy. Nonetheless, on the scale of continents drifting thousands of kilometers, there is no denying that paleomagnetism has already provided overwhelming support for Wegener's supercontinent and the plate tectonic continental drift that the existence of Pangea would have required.

•  •  •

Scientists able to change their minds have a huge advantage over those who cannot. Familiar with the arguments against a hypothesis, they are well suited to assuage the concerns that they themselves had previously harbored. But in order to change one's mind, one must keep it open to new evidence or arguments. Tuzo Wilson, a Canadian geophysicist and geologist, did just that.

Before the plate tectonic revolution, as late as the late 1950s, Wilson was one of the major opponents of continental drift and mantle convection. Even though Wilson permitted continental drift to disperse the continents away from their undeniable position in Wegener's Pangea, he seemed to think that continental drift was only a relatively recent phenomenon and that it could not explain the more ancient rocks billions of years old exposed in the Canadian Shield where he worked. Like most geologists, Wilson was a staunch uniformitarian, a belief embraced by Charles Lyell, who many accept as the founder of geology. Remember we discussed the uniformitarian maxim earlier, that "the present is the key to the past." Having to apply different rules to different periods of Earth history bothered Wilson. Because in his mind the continental drift theory had to apply to all of Earth history, and he didn't see any evidence in the ancient rocks of Canada, he rejected the hypothesis on philosophical grounds.

When it came to mantle convection, Wilson argued there was simply no empirical evidence to suggest its existence. Arthur Holmes, the founder of radiometric dating, had proposed mantle convection as the mechanistic savior of Wegener's continental drift hypothesis. But since little evidence for mantle convection was actually available at that time, few geologists were convinced and Wilson was no exception. In the short time between the spring of 1960 and the fall of 1961, Wilson would radically reverse his positions on both continental drift and mantle convection and become a leading figure of the fledgling plate tectonic revolution. Wilson ultimately changed his mind under pressure from mounting paleomagnetic evidence for continental drift led by Ted Irving and others, as well as the embryonic conception of seafloor spreading that was developing.

Even before his philosophical pivot, Wilson was already a leading international figure in geology, thus rendering his change of heart impactful on building scientific consensus. In 1963 in *Nature*, Wilson gave himself (and others he hoped to convert) an alibi: "The question whether the continents have been fixed in approximately the same relative position since their creation, or whether they have moved has been debated for fifty years. Perhaps the reason that this has never been settled is that much more is known about the continents than about the ocean floors, where the decisive evidence probably lies."[17] By the time the magnetic stripes of the seafloor were mapped during the Cold War, Wilson was one of the first to make their connection to the budding hypothesis of seafloor spreading. Wilson later identified the cycle of the oceans and the role their creation and destruction plays in the breakup and assembly, respectively, of the continents.[18] What has come to be known as the Wilson cycle should not be conflated with the supercontinent cycle itself, but the former turns out to be a critical cog in the latter, larger system.

• • •

For continents to collide, oceans must close; and for a supercontinent to break up, oceans must open. Ocean crust is therefore the currency that must be exchanged with the mantle in order to get continents to drift. Decades before the supercontinent cycle would be envisioned, Wilson already had a detailed vision of the life cycle of oceans. Although individual oceans are merely regional, when globally integrated, their collective life cycles in aggregate quite accurately describe the supercontinent cycle. In fact, the Wilson cycle is so integral to the supercontinent cycle that many people, even geologists, often conflate the two as the same process.

Puzzle pieces are an illustrative metaphor often used by geologists when describing the reorganization of the continents by plate tectonics. In fact, in most cases, this simplified thinking actually holds true. That is, the continents appear to largely act like coherent puzzle pieces over time, even after they have been reshuffled in different arrangements multiple times. Australia, North America, and Siberia are examples of continents that formed almost 2 billion years ago and yet still remain largely coherent continental entities today. Wilson was one of the first geologists to realize that continents tend to collide and break apart at largely the same margins, leading him to ask the question: "Did the Atlantic Ocean close and then re-open?"[19]

Even though Wilson himself was a geophysicist, it was fossil evidence that inspired his provocative hypothesis. Like Wegener before him, Wilson was willing to wade into the details of paleontology in order to advance his theories—theories for which geophysical evidence would be lacking for some time. There were two enigmas that Wilson thought he could resolve with one hypothesis. The first enigma was that on either side of the present-day Atlantic Ocean, the marine animal fossils of the Paleozoic era (541–251 million years ago) seemed to be remarkably similar despite their great separation. Wilson wasn't the first to propose that allowing for continental drift and the breakup of Pangea would allow these similar faunas to have been part of the same biogeographic range. Remember that one of Wegener's most compelling lines of evidence was his linking similar fossils found on continents on either side of the Atlantic Ocean.

The second and unresolved vexation was that on both sides of the Atlantic, there were localities in which dissimilar faunas lay adjacent to one another. Wilson's basic idea was that there had been an earlier Atlantic Ocean, a "proto-Atlantic" Ocean, for much of the Paleozoic era, and that it was the closure of this ocean that had juxtaposed the dissimilar faunas so that they were located side by side. This hypothesis provoked decades of research and we now know this hypothesized vanished ocean indeed existed and call it the Iapetus Ocean, mentioned earlier. In Greek mythology, Iapetus is the father of Atlas, just as the Iapetus Ocean can be considered a precursor to the Atlantic Ocean. Closure of the Iapetus Ocean is what gave rise to the ancient Appalachian and Caledonian mountain belts that can today be found on either side of the Atlantic Ocean, in both Atlan-

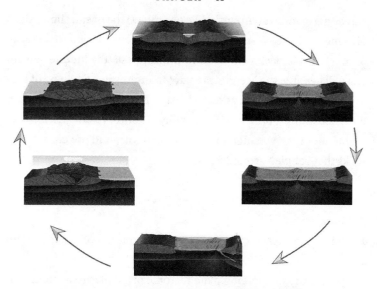

Figure 10. The phases of the Wilson cycle, starting from the top: (1) continental breakup, (2) opening of an oceanic basin, (3) seafloor spreading with widening of the basin, (4) subduction of oceanic lithosphere, (5) closure of the basin, and (6) continent-continent collision. Image reprinted from Wikipedia courtesy of Creative Commons license CC BY-SA 4.0: https://commons.wikimedia.org/wiki/File:Rock_cycle_in_Wilson_Cycle .png#/media/File:Rock_cycle_in_Wilson_Cycle.png.

tic Canada and Scotland. It was Wilson's resolution of the second enigma of the Atlantic fauna that distinguished his idea of a tectonic cycle.

What has come to be known as the Wilson cycle spans the entire life cycle of an ocean (fig. 10). Only two years after tentatively asking his provocative question, Wilson laid out the well-established six stages of an ocean's lifetime, using examples from modern tectonic settings to illustrate each stage of his idea.[20] In so doing, he was following the principle of uniformitarianism: "the present is the key to the past." These stages are: (1) Continents rift apart (think the East African Rift currently pulling apart new chasms every year); (2) a young ocean is born (like the shallow Red Sea, with parts even still hoisted above sea level); (3) the ocean matures (e.g., seafloor spreading leads to an Atlantic-like ocean); (4) the ocean shows signs of aging (as in the West Pacific, where the ocean crust is anomalously old); (5) the old ocean enters its terminal phase (the Mediterranean Sea will become the Mediterranean Mountains in a few tens of million years as Arabia creeps ever closer to Eurasia); and (6) the ocean dies, becoming a buried relic by being either recycled back into the mantle or preserved high in the mountains around the continental rocks that collided around it.

•   •   •

Plate tectonics played a key role in the formation of the mountain belts that antedated Pangea. Wilson's model, therefore, opened up the possibility of applying plate tectonic principles to ancient orogenic belts. ("Orogen" is the name geologists give old mountain belts that may no longer have a topographic expression due to weathering and erosion.) The questions of what, when, and where have largely been settled for Pangea. In technical parlance, the sum of these pieces of information adds up to what is referred to as "kinematics." The kinematics of the assembly and breakup of supercontinent Pangea are now quite well understood. We know what continents moved where and when. We also know the Wilson cycle implies that orogenic belts within Pangea are sites where ancient oceans opened and closed and then reopened to arrive at the kinematics, or the dance, of Pangea and the continents we have today. It turns out that a supercontinent neither assembles nor breaks up overnight, or even all at once.

Rome wasn't built in a day, and neither was Pangea. Rome didn't collapse in a day either, and neither did Pangea. Both supercontinent assembly and breakup typically happen one block at a time. Assembly involves the collision of continental blocks, until one, two, or only a few large blocks are left. Breakup involves rifting and the creation of smaller continents until the continents are totally fragmented and drifting individually. In this way, the supercontinent cycle can perhaps be most accurately described as the evolution from many plates to very few, and back again.

There are fundamental differences between the assembly and the breakup of Pangea. The puzzle-piece metaphor would seem to imply that continental pieces during supercontinent assembly and breakup are the same, merely reshuffled. But that isn't quite right. Take for example Laurentia, the ancestral North American continent that, since its formation almost 2 billion years ago, had included Greenland until 70 million years ago. Laurentia's collision with eastern Europe—resulting in Laurussia—occurred during supercontinent assembly; whereas the much larger and younger Laurasia, which included Laurussia, Siberia, and Asia, was the result of supercontinent breakup (fig. 11).[21] But this variability in the size of puzzle pieces during assembly and breakup is not by any means always the case. The fact that continents like Laurentia have persisted for so long, nonetheless, is a testament to the fact that rifts often open up along the same lines as prior collisional mountain belts since these are mechanically weak compared to the rigid continental interiors. A more accurate metaphor is a puzzle where the pieces themselves are growing and shrinking.

As discussed earlier, the first building block of Pangea to form was Gondwana, and even Gondwana itself was composed of several building blocks. Du Toit's Gondwana, comprised of all the continents of the southern hemisphere, had two halves: West and East Gondwana, which themselves formed from the amalgamation of smaller continents. Hundreds of millions of years before the rest of Pangea assembled, the assembly of Gondwana occurred across one of the most fundamental transitions in Earth history: the Precambrian-Cambrian boundary about 541 million years ago. West Gondwana formed immediately before the boundary, and East Gondwana formed immediately after the boundary, implying that the assembly of Gondwana straddling this transition coincided with

### Laurussia ca. 270 million years ago

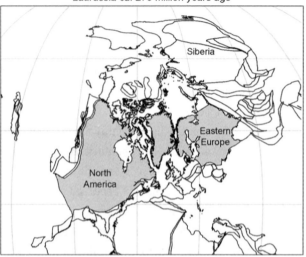

### Laurasia ca. 130 million years ago

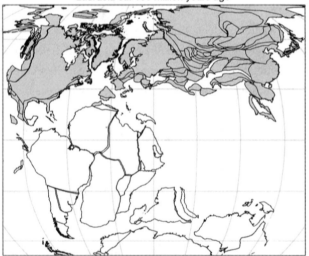

Figure 11. (*Top*) Laurussia about 270 million years ago, comprising North America (including Greenland) and Eastern Europe but not yet including Siberia (to the northeast) or various Chinese blocks (to the east, not shown). (*Bottom*) Laurasia about 130 million years ago during Pangea breakup, including Siberia and Chinese blocks.

(and may have had a profound effect on) this pivotal interval of irreversible global change.

The Cambrian "explosion" of animal life, largely occurring between about 530 and 515 million years ago, was not merely a profound radiation in animal diversity. Perhaps what distinguishes the Cambrian explosion the most are two unique innovations: the increase in disparity (types of body plans) and biomineralization (minerals, like calcite and phosphate, made by and for life).[22] Before this time interval, all macroscopic life was soft-bodied and so poorly preserved in the geological record. The assembly of Gondwana could have very well contributed to both these specialties of the Cambrian explosion. The "Pan-African" Orogen is the term given to the series of continental collisions that resulted in the assembly of Gondwana, and it is one of the largest mountain ranges to form in all of Earth history. However, this pan-African event also occurred at a pivotal time, when atmospheric oxygen was high enough for multicellular life to flourish. Both the size and timing of the pan-African collisions were primed to have just the right effect on the proliferation of different life-forms.

The taller the mountains, and the longer the mountain chain, the more the weathered material that makes its way as runoff into the oceans. Since animal life at this point in the dawn of the Paleozoic era had yet to crawl onto land, it was in the oceans that multicellular life took hold. The extensive flushing of nutrients and dissolved ions from weathered rocks into the oceans would have therefore had a maximum impact on marine animal life. The Cambrian explosion is the first time that animals made hard body parts. The innovation of widespread "biominerals," such as calcium carbonate or phosphate shells, requires critical elements like calcium and phosphorus, respectively, to be "bio-available." The weathering of Gondwana and its surge of nutrients and ions into the oceans could not have been better timed. Multicellular life, already benefiting from elevated oxygen levels achieved in latest Precambrian time, took advantage of a sudden abundance of dissolved goodies in the ocean.

The expansion of marine life due to the availability of critical mineral-making ions dissolved in the oceans may have further facilitated the rapid increase in the types of body shapes that distinguished the Cambrian explosion. If life can suddenly flourish in the oceans by making skeletons, it follows that a variety of fundamentally different ways of life could flourish at the same time. Disparity is different from diversity. For example, an ecosystem with one species of grass, one rabbit, and a hawk would have very low diversity but very high disparity. Because Gondwana would have facilitated a vast expansion in underwater *ways* of life, it could have spurred increased diversity *and* disparity. Life-made minerals—biomineralization—enabled not merely a way of making new types of life but also a new way of life.

The assembly of megacontinent Gondwana at the start of the Paleozoic era might have spurred on the largest evolutionary advance in history, but assembly of a supercontinent might have caused the largest mass extinction. Scientists believe the so-called P/T (Permian/Triassic) extinction, around 251 million years ago, was the most lethal of all mass extinctions, eliminating over half the animal species living at that time—twice as lethal as the mass extinction event that killed the dinosaurs.

Peter Ward and Joseph Kirschvink, two colleagues of mine and the authors of *A New History of Life*, argued in 2000 in *Science* that the P/T extinction appeared to have two tempos: a long decline in species leading up to the extinction boundary, combined with a sudden drop in animal diversity right at the boundary.[23] The sudden drop in species is almost unanimously attributed to the volcanic eruptions of the massive Siberian Traps and their deleterious effects on ocean chemistry. The cause of the gradual decline in species presaging the extinction, however, is less understood. But simply plotting the

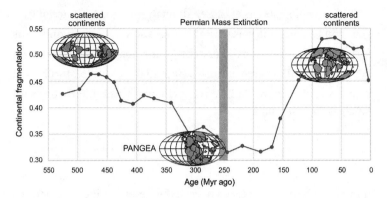

Figure 12. The formation of Pangea coincides with the most lethal of mass extinctions, the end Permian mass extinction. The degree of continental fragmentation is a nice way to plot with data the assembly and breakup of Pangea, which has been correlated with animal diversity. Adapted from A. Zaffos, S. Finnegan, and S. E. Peters, "Plate tectonic regulation of global marine animal diversity," *Proceedings of the National Academy of Sciences* 114, no. 22 (2017): 5653–58.

assembly of supercontinent Pangea leading up to the major extinction yields an intriguing correlation that is hard to ignore (fig. 12).[24]

How could assembling a supercontinent lead to extinction? The impressive diversity of Darwin's fifteen species of finches is thanks to the geographic isolation of the Galápagos Islands. One can think of the final assembly of supercontinent Pangea as creating habitat the exact opposite of the Galápagos. Remove the physical barriers to geographic isolation and then interbreeding increases, genetic isolation decreases, and diversity drops. If this happens on a global scale, diversity could conceivably drop catastrophically, as it did at this time. The converse evidence for this idea is also clear in the breakup of Pangea: as the continents have dispersed and geographic populations have become isolated (now akin to Darwin's finches), animal diversity has surged back over the last 250 million years (fig. 12). Life and death may be intimately related to the breakup and assembly of Pangea.

Today, geologists largely agree about the kinematic dance of Pangea. Thus having answered *what*, *where*, and *when* supercontinents form, we can now begin to ponder *why* and *how* the supercontinent cycle exists at all. And so we have returned full circle to Murphy and Nance's conundrum: Can we reconcile plate tectonic theory with how Pangea actually formed?

·   ·   ·

If there's any personal stake I have in the supercontinent debate, it's that there's a critical ingredient missing from existing models: the mantle. This is a big omission. Earth is a stratified layer cake and the mantle is the thickest layer, approximating half of Earth's radius and 84% of its volume. By mass, the mantle comprises over half the planet. The mantle is also the go-between Earth's core, which stores primordial heat, and the crust that releases it. Remember it is internal heat that fuels plate tectonic convection. It is thus mantle convection that regulates the speed at which the Earth engine burns through its fuel supply by transferring heat to the crust. It is very likely that the supercontinent cycle represents the long-term overturning of the mantle—understand the mantle and its convective cycles, and you understand where the supercontinent cycle is taking us. Also recall that plate motions are intimately linked to the upwellings and downwellings of mantle convection, and were germane to the Pangea conundrum—if we understand the mantle better, we may get closer to the paradigm shift Murphy and Nance were seeking.

How can we study the dark depths of the mantle? Seismology. As devastating as earthquakes are to our way of life, they provide an image of the underworld much like an X-ray provides an image of a patient's internal organs. Earthquakes release seismic energy in the form of waves that ripple all around the world, and even all the way *through* the world. Those waves that ripple along Earth's surface are called (unimaginatively) "surface waves." While surface waves are

informative in their own way, for our purposes we are more interested in those waves that permeate the depths of the earth, so-called
body waves. A seismometer, the instrument we use to detect seismic
waves released by earthquakes, has a needle that bounces when the
waves finally arrive. There are two types of body waves that arrive at
different times. P-waves are the fastest and arrive first ("primary").
S-waves arrive second ("secondary"). Incidentally, but conveniently,
P-waves are also called that because they are "pressure waves" and
S-waves also have an alternative namesake because they are "shear
waves." It is due to their different means of travel that P- and S-waves
arrive at different times. And it is the slower S-waves whose shear-
ing motion tells us the most about the deepest regions of the mantle.

The speed of shear waves is dependent on the nature of the material through which they pass, which is why they tell us so much about
inner Earth. When shear waves are slow, this is due to either elevated
temperature or chemically distinct material. Studying these waves
led to the identification of buried scientific treasure in the form of
two massive blobs sitting at the base of the mantle on either side of
the world. One blob is located beneath the African plate and another
at the opposite end of the Earth, underneath the Pacific Ocean (fig.
13). These blobs can be detected due to their slow shearwave speeds.
Thus, these massive blobs must be either thermally hot or chemically buoyant, and are most likely both hot and buoyant.

What I am calling "blobs" seismologists call by such an egregious
abbreviation I only perpetuate it reluctantly: LLSVPs, for "large low
shearwave velocity provinces."[25] It is a technically apt name, but it's
no wonder you probably haven't heard of LLSVPs, as important as
they are to completing the plate tectonic revolution. But geologists
waffling over a snappier name (such as myself) must be ignored
for the moment because we really are uncertain of the origins of
the LLSVPs. If we called them "piles," then we'd be implying some
mechanism of piling things up, which may lend less credence to
their thermal properties and more to their composition.

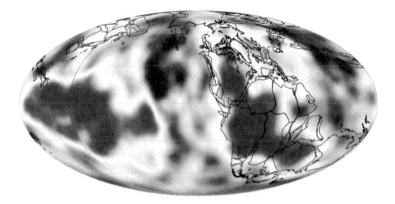

Figure 13. Mirror image of lower mantle LLSVPs (light gray under both Pangea and the Pacific Ocean) and Pangea about 200 million years ago. African LLSVP (*right*) and Pacific LLSVP (*left*). Light gray are slow and dark gray (in between Pangea and the Pacific) are fast present-day seismic velocities imaged at the core-mantle boundary (~1,740 miles or ~2,800 km depth). Note the deep African LLSVP closely matches the shape of supercontinent Pangea at the surface.

The size of both these deep cryptic bodies is vast. Without hyperbole, these deep mantle LLSVPs are the biggest geologic features we have on the planet today, bigger than any given continent, even the megacontinent of Eurasia. And their origins may be quite old. Some of the material swept up in these LLSVPs may be as old as the planet. Before Earth had any buoyant continents made of granite, its surface is thought to have been just a bubbling magma ocean. During Earth's first few fiery millions of years this magma ocean started crystallizing. Light elements (like silicon) floated, dense elements (like iron) sank. This is why the mantle of the planet is a silicate shell. But this is also maybe why there may be dense iron-rich particles sitting at the very base of the mantle. But what do these dense LLSVPs in the lower mantle have to do with supercontinents?

Simply put, the lower mantle LLSVP beneath the African plate is the shape of Pangea! Yes, you heard that right, the dense particles sitting at the base of the lower mantle reflect like a mirror the shape of

Pangea at Earth's surface, and this is by no means a coincidence. We know this because we can reconstruct the ancient position of Pangea, and multiple independent methods all show the same reflection: the African LLSVP sitting today in the lower mantle retains the location and shape of Pangea 200 million years ago.[26] Now, this particular discovery involves solving a technical but critical problem to which we will have to return later: the paleolongitude problem. Just as in navigation, longitude has been an elusive issue in reconstructing the positions of continents. For now, suffice it to say that the remarkable correspondence between the LLSVP in the lower mantle and Pangea indicates we're on the right track with solving paleolongitude and have a real clue on our hands. But what does it mean?

To be clear, it is a mirror image reflected across time. That is, the *present* lower mantle reflects the *ancient* shape and position of Pangea. This time lag between the surface and the deep can be explained by several phenomena. First, subduction of ocean crust takes millions of years, in fact *hundreds* of millions of years to descend through the entire thickness of the mantle. If the dense particles in the lower mantle are swept together by subducting slabs of ocean crust, then these blobs would take a long time to develop. That is, the picture in the lower mantle is likely to reflect the picture of the surface from quite some time ago. Second, once Pangea took its final shape, it was surrounded by subduction zones at its edges, but in this respect, little has changed in the ensuing 300 million years. Even as the supercontinent started to break up roughly 200 million years ago, these fringing subduction zones have stayed with their dispersing continents, allowing their drift to be accommodated by the consumption of ocean crust in their paths. Think of South America and the Andes being thrust up, or North America's Rocky Mountains. Although we give these mountain chains different names (the "Canadian Rockies" in Canada, of course), they are part of the same system: the Cordillera, which is the mountain belt thrust skyward due to ongoing subduction at the edge of the continents drifting *away* from Pangea.

The point being, subduction has not disturbed the lower mantle beneath Africa, the center of Pangea, for quite some time. Just like buried treasure, the lower mantle LLSVP beneath Africa, reflecting the ancient shape and position of Pangea, has been waiting a long time to be found, and, thanks to modern seismology, we have done just that.

But what does the LLSVP beneath the Pacific plate have to do with Pangea? Speaking of mirror images, the two LLSVPs are very much mirror images of each other. Sure, there are slight differences, but it's their many similarities that are most striking: they are equally massive and both straddle the equator. And to make our mirror image analogy even more accurate, they are reflections of each other. Take a journey through the center of the earth from the center of the African LLSVP and you come out on the other side in the center of the Pacific LLSVP. This equal and opposite organization of the deep mantle can only be explained by two massive mantle convection cells. These two LLSVPs are evidence that convection in the mantle is so large that it can be characterized by only two cells: two hot upwellings beneath Africa and the Pacific, with a ring of cold downwelling that bisects them. This downwelling is the same ring as the ring of fire (fig. 8). Thus, plate tectonic subduction occurring along the ring of fire is in lockstep with whole mantle convection. Plate tectonics and mantle convection are so intrinsically linked, we cannot understand the surface without looking and thinking deep.

Most importantly, the LLSVPs might help us resolve the Pangea conundrum (fig. 2). Murphy and Nance had assumed that the young Iapetus and Rheic Oceans opening up internally to the continents had resided over a mantle upwelling and that the continents were drifting topographically downhill to regions of mantle downwelling. They even saw evidence for "intra-oceanic" subduction occurring within the paleo-Pacific Ocean that they took to be corroborating evidence that the continents were planning to consume the old, external ocean as they drifted toward the mantle downwelling away from the upwelling from which they had come. This led them to the conundrum that if that were the case, then Pangea should have formed inside-out.

But as we know, that didn't happen, and the continents drifting toward the paleo-Pacific suddenly reversed course and instead consumed their young, internal oceans. Like any responsible scientists pointing out a problem, the duo ventured a guess as to its solution.

They nearly solved their own riddle and are worth quoting directly: "Assuming continents migrate from [topographic] highs to [topographic] lows, then the reversal in continental motions that led to the formation of Pangea may have coincided with the emergence of a [topographic] high in the paleo-Pacific [ca. 450 million years ago]."[27] Murphy and Nance realized that their assumptions about how the mantle was convecting beneath the opening and closing oceans might have been flawed. Continents might have been drifting downhill toward a mantle downwelling under the paleo-Pacific Ocean, but what if this changed? What if hot mantle started upwelling under the paleo-Pacific just as the continents of Pangea reversed their course and collapsed back in on themselves, forming Pangea the way we know it did? Do we have any evidence of this?

Yes. And the LLSVPs are just that. Although seismic and geophysical data since the 1980s had foreshadowed their discovery, the two large low shearwave velocity provinces were not formally identified and given their egregious abbreviation until 2007—only the year before Murphy and Nance wrote their paper. As with most new discoveries in science, the LLSVPs were debated in the secluded halls of specialists before the rest of the field started catching on. Even if science is ideally like a crossword puzzle, these interdisciplinary cross-checks between acrosses and downs are typically conducted in the long term. Before geologists like Murphy and Nance could rely on the LLSVPs as corroborating evidence, they not only had to be aware of them, but they also had to be assured by the seismologists of their validity.

But even if Murphy and Nance had managed to catch wind of these contemporary advances in seismology, it still took a few more years before geologists began to get an idea of the ages of the LLSVPs. After all, the LLSVPs are *presently* sitting in the lower mantle. The fact that the African LLSVP matches the shape and position of Pangea proves that it is likely at least as old as Pangea. Thus it is generally accepted that the *present-day* LLSVPs are actually at least 200–300 million years old. Recall that some of the dense particles piled up in these LLSVPs at the base of the mantle could be almost the age of the Earth, and indeed some geologists think these LLSVPs could be billions of years old. Such antiquity of the LLSVPs, at least as they are piled up in their present positions, is much debated.

But for the Pacific LLSVP to have been the savior of the Pangea conundrum, it only has to be as old as the confusing reversal of plate motion that closed the young, internal oceans to create Pangea. The surprising reversal of plate motion was a two-step process that required subduction of both internal oceans, the Iapetus and the Rheic. First, subduction of the Iapetus Ocean began about 480 million years ago, but its neighboring internal ocean, the Rheic, was still spreading. Then, when subduction of the Rheic Ocean began about 430 million years ago, the reversal of plate motion was complete. The hot mantle upwelling in the paleo-Pacific thus would have to be approximately 500 million years old, roughly twice the age of the African upwelling beneath Pangea (fig. 14). Furthermore, even if mantle upwelling in the paleo-Pacific initiated early enough to explain the reversal of plate motion, its position relative to the continents also had to make sense with a thermally uplifted mantle pushing the originally westward continents back to the east.

First, the test of its age. Murphy and Nance had already conducted this due diligence. Even when scientists speculate, there is an obligation to mention any basic logic or obvious evidence that could be used to test it. Indeed, Murphy and Nance noted there was evidence of the mantle "superplume" they envisioned pushing continents back

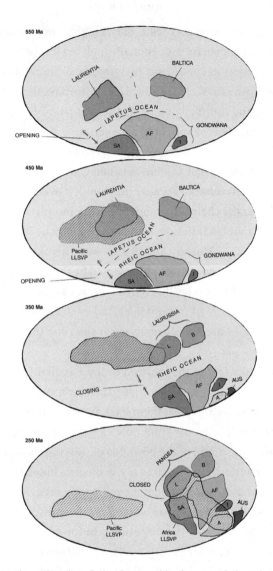

Figure 14. Continental motions during the assembly of Pangea relative to lower mantle superplumes (LLSVPs). The opening of the Iapetus and Rheic Oceans pushed Laurentia (North America) and Baltica (Europe and eastern Russia) away from Gondwana, but then, as Laurentia encountered the topographic high of the newly developed Pacific superplume, its motion reversed. Explaining this tectonic reversal—which ended up closing these young, internal oceans to create Pangea—thus resolves the Pangea conundrum. See figure 2 for abbreviations.

from traversing the paleo-Pacific. Most notably, the period of time in question was known for exceptionally high sea levels, flooding well into the interiors of continents. Imagine ocean shorelines reaching Minnesota. An easy way to raise sea level is to make the ocean basins shallower, and thermally uplifting the vast paleo-Pacific seafloor with a mantle superplume could have certainly achieved this.

Recall how I stressed giving equal weight to both LLSVPs' thermal and chemical origins? Many seismologists don't like calling the LLSVPs "superplumes," because they think it overemphasizes the thermal over the chemical origin. But in the interest of our present discussion about mantle convection, the hypothetical paleo-Pacific superplume that Murphy and Nance were invoking could have indeed been real and possibly represented by the Pacific LLSVP that we still have with us today (fig. 13). (It's always a kick when scientists theorize about something and later find evidence already existed of which they were simply unaware at the time they wrote their paper.)

Second, the test of the location of the Pacific LLSVP. Was the Pacific LLSVP in the correct position to reverse the westward motion of the continents and push them back from whence they came? I don't want to lose momentum now explaining the technicalities of how we can be sure of the positions of the continents relative to the deep mantle, but suffice it to say, geologists indeed have means to do so and these independent means largely seem to agree. If we couldn't do this precisely, then we wouldn't have discovered that the African LLSVP matches the shape and position of Pangea about 200 million years ago, as in figure 13! So please just trust me for now. Back to the final test. Indeed, the Pacific LLSVP was perfectly positioned to push back against the young oceans opening at that time. The Pangea conundrum can thus finally be resolved by the Iapetus and Rheic Oceans opening up until North America (Laurentia) encountered the developing Pacific LLSVP and was pushed back toward Gondwana in a surprising reversal of plate motion (fig. 14).

In this clash of tectonic forces, the push from the superheated Pacific mantle bested the push from the ridges of the young interior oceans (fig. 14). The Pangean continents thus collapsed inward, creating the supercontinent we know existed. Murphy and Nance may have emphasized the conundrum, but in so doing, they also helped themselves and others think creatively about which of their assumptions potentially needed to be discarded. As it turns out, the patterns of mantle convection beneath the oceans were different than they had initially assumed. But even in their own speculation, only after having identified the paradox of course, were they able to grasp at the solution we now hold more firmly today. As a last-ditch effort, the duo had guessed that the continents had been drifting toward a topographic low but might have encountered a topographic high along the way. And indeed, the continents did. And the evidence for that topographic high, the Pacific LLSVP, was only discovered a year before Murphy and Nance wrote their paper.

But is it really feasible that the Pacific LLSVP may be nearly twice as old as its mirror image, the African LLSVP? Not only is it feasible, but it is in fact required. Ironically, even as the great antiquity of the Pacific LLSVP helps resolve Murphy and Nance's Pangea conundrum, if the African LLSVP were just as old as its Pacific counterpart (i.e., ~500 million years old), then we would just be creating a whole new Pangea conundrum. Recall that Pangea fits snugly over the African LLSVP (fig. 13), but Pangea is only about 320 million years old. If the African LLSVP already existed, how could Pangea have formed over a mantle upwelling? Pangea would have had to climb topographically uphill, which we have just argued is physically implausible for drifting continents to do. That is, if the reason Pangea didn't form over the paleo-Pacific Ocean is because the Pacific LLSVP pushed the continents back, then we couldn't explain how Pangea then formed above the African LLSVP—if indeed the African mantle upwelling already existed. Thus, the only way to avoid creating a new Pangea conundrum by following the logic that helped us solve the last one, is if we assert that the African LLSVP didn't yet exist when Pangea formed.

But such an ad hoc explanation of such an important matter is less than satisfying. Luckily, there are a few additional angles we can take to test this idea that the Pacific and African LLSVPs are significantly different ages. Unfortunately, studying Pangea isn't enough to reveal all the mantle's secrets. If the African LLSVP indeed postdates the assembly of Pangea, then it is likely that Pangea somehow formed the African LLSVP, possibly by forming a ring of fire of subduction around it and maintaining it. But did Pangea in fact assemble on the opposite side of the world from the Pacific LLSVP? And if so, why? In order to understand these buried mantle treasures, we need to understand where Pangea came from, and thus what came before it.

# 2
# RODINIA

I lost half my right thumb for Rodinia.

It was the summer of 2006 and I was an undergraduate field assistant for Nicholas Swanson-Hysell, who was conducting field work for his PhD at Princeton University. Academia has its roots in medieval guilds and little has changed since. Students have mentors, and Swanson-Hysell, in my eyes, was certainly my mentor. The truth was he also had his mentor, Adam Maloof, his professor now successfully tenured at Princeton. But back then, we were all young and hungry.

I say Swanson-Hysell was my mentor for many reasons, but for this reason perhaps above all. His mentor, then Assistant Professor Maloof, trusted Swanson-Hysell implicitly and would entrust the two of us—just Swanson-Hysell and me—to conduct the well-planned weeks ahead doing fieldwork in the Australian outback. I said summer, but this is my northern hemisphere bias speaking. Yes, it was boreal summer, but it was winter in Australia—austral winter. No one would dare do fieldwork in the outback during austral summer unless they had a death wish. Winter is hot enough down under.

I couldn't believe my luck to be there. Even though Swanson-Hysell and I had known each other as classmates from Carleton College, Professor Maloof still made me apply for the coveted job. There were of course the academic course records—what classes had I taken and how well had I done? But I also had to answer the question: "Why do you want to hit the outback?" I guess Maloof was testing how much I knew of the backstory for why he and Swanson-Hysell were working in the remote outback of all places.

How little I knew at time. And how easily I was reminded of this as soon as the trip got underway. The trip started, for me at least, with what geologists like to call "geotourism." Even as single-minded as scientists are in the pursuit of collecting the exact samples they need to test their hypotheses, since they've typically traveled so far, it would be a shame to not save a few days in the beginning, but more typically at the end, to see the sights and explore the best-known or best-preserved rocks in the area that *aren't* the exact ones they'll be spending most of their waking hours sampling for the next few weeks or months—like tacking some tourism onto a business trip, except the business is rocks and the tourism is rocks.

We were in South Australia in the outback north of Adelaide and the Clare Valley wine country. The scenic Flinders Ranges National Park, with some of Earth's first macroscopic forms of life, made for ideal geotourism. And even though Swanson-Hysell and I alone would take up the reins of conducting the fieldwork for his PhD, the whole crew was still there. Now tenured professors at McGill and Northwestern Universities, Galen Halverson and Matthew Hurtgen were close colleagues of Maloof's and they had all come up the ranks together. Halverson and Maloof had done their PhDs with Paul Hoffman—the giant in our field of geology (fig. 15). So naturally the campfire the first night featured stories about Hoffman. Such is academic folklore, passed down from generation to generation. Swanson-Hysell and I listened intently as the smokeless fire of the gumtree crackled.

Perhaps the moral of the first story they told was that they would be more forgiving to me as a new student than Hoffman had been to them. In the field, what you know and don't know is immediately on utter display. There's no way to hide at a rock outcrop. As soon as geologists descend on an outcropping of exposed rock, they simply

Figure 15. Paul Hoffman in Tillite Gorge of the Flinders Range of South Australia. He is standing on the glacial deposits that he made famous with the Snowball Earth hypothesis. Photo credit: Adam Nordsvan.

can't contain their excitement, sharing every observation they have made or uncertainty they may have. And after a back-and-forth discussion of trial and error, it is as a group that a team of geologists will come to a consensus interpretation of what the edifice of rock tells us—or at least, we know what we'll be arguing about over the crackling fire that night!

At the time of their story, Maloof and Halverson were PhD students of Hoffman. It was one of their first times in the field with their new mentor. Like us, they were eager to impress, eager for any indication of approval—and terrified to say something stupid. But if you say nothing at all, that's worse. At an outcrop with others, sure one can and must think pensively, but it is also an opportunity to have a pointed discussion where hypothesis and hypothesis testing play out in real time like a rapid-fire exchange in our own patois: rock talk.

Rocks are composed of minerals. If one wants to correctly identify a rock, it is thus wise to start with correctly identifying some minerals. But unlike large rocks that may either form cliffs or fit in the palm of your hand, minerals are usually too small to see clearly with the naked eye, which is why the hand lens is an essential part of a geologist's field gear. How good you are at identifying minerals with your magnifying glass often decides your fate as an outcrop geologist. Sure, there are many other larger-scale features to glean from an outcrop, but if you can't identify the rock, then everything else is a bit dubious.

Mineralogy is one of the first upper-level classes students of geology take. Minerals are everywhere. They are mined to make the materials in our cars and phones, they are in lipstick, they are even in our bones. There is in fact no definition of a rock other than an aggregate of minerals. It is these shimmering and unique crystals in natural history museum gift shops that are to blame for turning many young people into geologists long before they know it. At least as geologists define them, minerals are crystalline compounds with diagnostic chemical composition and crystal structure. And since we are unable to conduct X-ray diffraction analyses with our eyes in order to glean the elements of their composition, it is usually the crystal structure that geologists are most looking for under the up-close gaze of their hand lenses in order to identify a mineral. A mineral's external appearance is indicative of its internal structure . . . so in this sense you can "judge the book by its cover"!

Hoffman had his students looking at a sedimentary rock, and it was that moment when, having already sized up the rock outcrop on the approach, they were ready to get on their hands and knees to see what their hand lenses would reveal. To use a hand lens correctly, one must get all the relative distances and the angle of the light just right. This involves getting both your eye close to the lens and the lens close to the rock until you find the right focus in the right light. It is a hushed moment. With one eye closed and face up against the outcrop, geologists can't help but announce the minerals they see, as they see them.

"Pyrite?" Maloof both said and asked, offering a mineral identification to his new mentor.

"Pyrite??" Hoffman echoed, but amplifying Maloof's uncertainty, and with a hint of exasperation.

"Pyrite???!?" Hoffman repeated after a dramatic pause, this time making his exasperation clear. This was now a rhetorical question, and Maloof, dejected, realized the answer was no, it wasn't pyrite.

"Well, I see the cubic crystal structure . . ." Maloof offered with hesitation. Having realized he'd made a mistake, he hoped to save face by offering the logic behind his educated, but wrong, guess.

"Sure, fine, cubic structure, but other minerals have cubic structure, and what about geologic *context*? What does that tell you?" Hoffman, even in his exasperation, continued to teach.

Pyrite, a.k.a. "fool's gold," is an iron sulphide ($FeS_2$). It most often occurs in igneous or metamorphic rocks, less often in sedimentary rocks. Because pyrite decomposes when exposed to oxygen, it is in fact quite rare to find pyrite in sedimentary rocks that form by weathering and deposition at Earth's oxygen-rich surface. Certain sediments that are deposited in stratified, poorly mixed water (such as the Black Sea) can allow for sedimentary deposition of pyrite, but this is a special case, and it wasn't the case in front of then-student Maloof.

The mineral that they were looking at was actually halite ($NaCl$), or simply rock salt. To Maloof's credit, although halite and pyrite have different chemical compositions, they both have a cubic crystal structure—recall that this is often the best clue a geologist has when identifying minerals. But there was a reason Hoffman made this a big deal and a teachable moment, and there is a reason Maloof still tells the story to this very day, even as self-deprecating as it may be.

Part of the story being told, at least from my perspective as a green-eared undergraduate at the time, was that it's okay to be wrong as a student. In fact, that is the best way to learn. Once a geologist sees something for the first time, they may get its interpretation wrong the first time, but never again. Once observed, a field observation is always there when you need it, at a moment's notice. My undergraduate advisor, Cameron Davidson, always told me, "She who sees the most rocks, wins." And how right that was. Becoming a seasoned field geologist is about trial and error. And although Maloof might have made a mistake that day, he surely never made it again.

Another part of the story was for the young professors to reminisce and to share with the next generation (Swanson-Hysell and me) the personality of their great mentor. Hoffman might have been a bit heavy-handed in his mentorship—the "back in our times" aspect provided the comedic relief of the campfire story—but the truth is that some scientists take few things more personally than their science. And scientific mentorship for them is a matter of honor.

Hoffman was teaching the importance of the whole outcrop. If Maloof had considered the larger context of the rock outcrop—the rock that the cubic mineral was hosted in—he not only could have ruled out pyrite as an option, but he could have even more importantly suspected halite even before it came into focus under his hand lens. The sedimentary rock they were looking at was an "evaporite," a salt deposit left after the evaporation of water. Even before they saw the salt minerals themselves up close, they had been ambling over mud cracks: desiccation features that indicated that water was in vanishing supply when these sediments had been deposited.

The lesson is thus that context matters, and we need to look at all scales for evidence. This is why I am telling this story about hand lenses and minerals in a book about massive supercontinents. Supercontinents are composed of their rifted blocks, known as continents. Continents are composed of rocks, and rocks are composed of minerals. Constraints on either scale, large or small, are ignored at our peril. If one is interested in the small mineralogical scale, then the drift of continents into the doldrums, or "horse latitudes" where winds are slow and there is little precipitation, is a great recipe for evaporation and forming an evaporite salt deposit. If one is interested in the large continental scale, then the identification of cubic halite crystals in an evaporite deposit is an invaluable constraint on the latitude of the continent at the time the sediment was deposited. Remember our scientific crossword puzzle? One can think of the large and small scales as internal cross-checks for the collective hypothesis being tested. Furthermore, the concepts and characters mentioned in this little anecdote will all play large roles in the story of supercontinent Rodinia.

My own story about losing half my right thumb for Rodinia will have to wait for its proper place in the narrative. The story of Rodinia doesn't have an equivalent of Pangea pioneer Alfred Wegener. Rodinia, from the Russian родина, *rodina*, meaning "motherland, birthplace," started with the conception that Pangea wasn't alone—that there might have been a supercontinent in even more ancient times before it. This was arguably a collective realization by the geologic community more than by any single scientist.

•   •   •

The trick with testing for pre-Pangea supercontinents is that many of the rules change. Most of the most compelling lines of evidence we used to reconstruct the continental configurations of Pangea simply aren't available. The diagnostic fossils that Wegener had used to correlate continents across the Atlantic Ocean are not available—multicellular life was only beginning to evolve during the ancient times of Rodinia. The fossils preserved during the age of Rodinia are typically microscopic and soft-bodied, and so are poorly preserved, so are not easy to find, and so are not easy to identify once you've found them. The first macroscopic fossils large enough to find at an outcrop (that is, without the aid of a microscope back at the lab) don't appear until about 600 million years ago. The fossil record improved exponentially when life-forms developed hard exteriors, some 540 million years ago. Although we will get to know more about these earlier soft-bodied creatures and what they tell us about the coevolution of life and the continents through history, they are not as helpful in the correlation of continents as their descendants were for Pangea.

The lack of the reliable lines of evidence we used to reconstruct Pangea during Rodinia times doesn't stop there. Recall that the post-Wegener plate tectonic revolution relied heavily on realizing the treasures under the sea, namely the modern seafloor. The complete recording of seafloor spreading during the breakup of Pangea is beautifully preserved on the present seafloor. The currently active spreading ridges that are still pushing the continents apart to this day, for example the Mid-Atlantic Ridge, all sit squarely in the middle of the oceans. From the active spreading ridge toward the continents, in either direction, the age of the ocean crust gets progressively older, with the oldest seafloor emplaced when Pangea first started breaking apart about 180 million years ago.

The plate tectonic revolution was stimulated by our knowledge of the crust hidden beneath the oceans, and since that time the global seafloor has been mapped thoroughly. As we discussed in the last chapter, the configuration of Pangea we get from this detailed record closely matches the Pangea configuration we get from deducing the ancient latitudes of the continents from their paleomagnetic record. In sum, both the oceanic and continental paleomagnetic records can be used to provide independent evidence for Pangea's existence.[1]

As I've also mentioned, ocean crust is an ephemeral feature at Earth's surface. Compared to the buoyant continents comprised of granite, the dense ocean crust made of basalt gets pushed back into the mantle when continents collide. The "internal" Atlantic-like oceans that opened up during the breakup of Rodinia are not beautifully preserved as they are for Pangea—they have all but vanished because of younger tectonic activities. As we will see, occasionally some fragments of ocean crust get scraped off onto a continent even as most the oceanic plate is subducted downward into the mantle.

While these fragmentary records of ancient oceans are clues nonetheless, they by no means resemble the impeccable records of seafloor spreading we used to decipher Pangea. The case for Rodinia must be made using different types of evidence.

The most conspicuous clues of ancient continental collisions are ancient mountain ranges. In the 1980s, clues like these began to suggest an earlier supercontinent, before Pangea. To be clear, ancient mountains don't look like they do today, and they aren't as easy to spot as towering Mount Everest. This is not only because of ceaseless erosion over hundreds of millions of years. In fact, once the tectonic forces that led to the collision of two continents have been overcome, or have been reorganized to focus their efforts elsewhere, a mountain belt will begin to collapse as the thickened mass at once towering heights is suddenly able to spread laterally. Such extension of the crust following the cessation of a continental collision is in fact possibly even more effective at razing a mountain range than erosion is. Either way, the combination of erosion from above and collapse from within serves to bring ancient mountains back down to earth, back toward their original thickness.

Geologists fortunately have clues other than topography to identify ancient mountains that signify the collision of continents like the modern Himalayas. And conveniently, it is ironically the erosion of ancient mountains (also called orogens, recall) that actually reveals these more resilient clues. As always in geology, the clues are in the rocks. So far we have mentioned both sedimentary rocks (e.g., Hoffman's disbelief over Maloof mistaking halite for pyrite in an evaporite deposit) and igneous rocks (e.g., the granites and basalts of the continental and oceanic crust, respectively, as well as the lavas on the seafloor and continents that record the ancient magnetic field so well with paleomagnetism). But the resilient clues of orogens hidden deep beneath the peaks of towering mountains comprise the third, last type of rock: metamorphic rock.

There is a good reason we have waited until now, after we have already acquainted ourselves with igneous and sedimentary rocks, before getting to know metamorphic rocks. As their name implies, metamorphic rocks have undergone metamorphosis, in this case, a transformation in minerals and textures. Every metamorphic rock started its life off as either a sedimentary or igneous rock. But even though metamorphic rocks may be derived from other rocks, they truly take on lives of their own. Sedimentary rocks form on Earth's surface and igneous rocks form either on the surface or relatively near the surface. Metamorphic rocks, however, are the result of what happens when tectonic forces take these surficial and shallow rocks and bring them to great depths.

Both temperature and pressure increase as you go deeper in the earth. But why? Earth contains great internal heat. This is why geothermal heat, when close enough to the surface like in Iceland and New Zealand, is an economically viable source of energy. And the increased heat encountered at these relatively shallow depths is nothing compared to deeper in the crust where most metamorphic rocks form. Without getting into details that we will explore later, the inner Earth has multiple sources of internal heating. One heat source is the "fossil heat" left over from Earth's fiery accretion as it formed as a planet 4.543 billion years ago. While this fossil heat, supplemented by radioactive heating, has been dissipating to space for a while now as the planet cools, we have learned that Earth still has significant reserves of internal heat remaining. Deep down, even after billions of years of cooling, Earth is still quite hot.

The trend of pressure increasing with depth is generally the same as it is for temperature, but for different reasons. Simply put, the more rock on top of you, the more pressure you are under. So how and why do rocks end up so deep? In a word: subduction. This is the mechanism that causes dense basaltic ocean crust to be recycled back into the mantle. It is also the mechanism that takes sedimentary and igneous rocks formed at or near the surface to potentially great depths where they are transformed into metamorphic rocks. Although this may be counterintuitive, mountains become high and deep at the same time, developing both towering peaks and deep roots. Both these seemingly opposed ends are achieved by the same simple means: thickening the crust. Unlike dense ocean crust that largely sinks back into the mantle, the thin skin of sedimentary rocks on top of it mostly gets scraped off. Instead of losing this sedimentary material to the mantle, it is added to the mass of the mountain range. Then, when so much ocean crust is consumed that the ocean between two continents vanishes, continents must collide. This is when the real mountain building—both up and down—begins.

Unlike vanishing ocean crust, less dense continental crust stays near the surface. Even if some crust is sent to great depths, it is too buoyant to be lost to the mantle. Whichever continent was *not* connected to the descending ocean crust gets the upper hand when the two continents eventually collide. The continent of the upper plate will always be placed above the continent attached to the subducting plate. So, even though India was moving very fast when it collided with a nearly stationary Eurasia, since India was embedded in a subducting oceanic plate, Eurasia got the upper hand during continental collision. Because of this, the crust of India has been pushed beneath that of Eurasia. So, at the same time that the Himalayan Mountains are being thrust up, a deep crustal root is also forming deep below. It is at these depths during such continent-continent collisions that metamorphic rocks form.

However, near-surface rocks are removed, exposing the deep inner guts of the former mountain chain. These rocks preserve visible evidence of the tectonic stresses involved in mountain building. Their minerals are those that are stable in the deep crust, and the rock layers, which were planar and horizontal when they were first deposited, are deformed into very complex geometries. Some minerals preserved in these rocks suggest they were exhumed from depths as much as 19 miles (30 km). Given that the highest mountain on Earth, Mount Everest, is less than 6 miles (10 km) high, such rocks are windows into the deep crust beneath modern mountains as well as physical evidence of the stresses involved in making mountains.

Long after an ancient mountain range has lost its craggy topography due to erosion and collapse, the evidence for its former existence is found in the metamorphic rocks it forged at great temperatures and depths. Not long after the plate tectonic revolution vindicated Wegener's Pangea hypothesis, the geologic community started to recognize that metamorphic rocks much older than Pangea seemed to indicate other periods of continent collisions—and the evidence was widespread enough to imply the existence of even older, pre-Pangean supercontinents.

• • •

The Appalachians are no Himalayas—or at least they're not anymore. The Appalachian Mountains were created some 350 million years ago when Pangea formed due to the collision of two large blocks: Laurentia (ancestral North America) and Europe, already connected since around 430 million years ago as "Laurussia," with Gondwana (which itself had already formed as early as 520 million years ago). Today, the Appalachian Trail traverses the entire spine of the long-since eroded and collapsed mountain belt. I do not wish to diminish the perseverance of committed "thru hikers" that trek the whole 2,190-mile-long (3,525 km) trail in one go—no one is conducting such a feat in the Himalayas—maybe geology will help future Earth inhabitants to do so in a few hundreds of millions of years! The point is, because the Appalachian Mountains are deeply eroded, the deep rocks have been uplifted. Therefore, the hills we hike over today are in fact the deeply exhumed root of the mountain belt, once tens of kilometers and miles underground but now exposed at the surface.

Although the Appalachians formed during the assembly of Pangea, there is an even older mountain belt nearby that was related to the assembly of previous supercontinent Rodinia. And thus even the now relatively smooth spine of the Appalachians is actually mountainous compared to this yet even older mountain belt. In fact, unless you were a geologist, and only if you were one who specializes in metamorphic rocks, would you even know you were standing on an ancient mountain range that once stood at towering heights.

Grenville is small Canadian town along the Ottawa River, between Montreal and Ottawa. This sleepy hamlet is the namesake of the Grenville Mountains that were to supercontinent Rodinia what the Appalachians were to Pangea—the formative collision. Although Grenville is home to a famous mountain belt, you certainly couldn't guess it from the flat topography of this now low-lying canal town. Nonetheless, there are clues in the rocks of the ancient mountains that once stood there. Just like the eroded Appalachians, the Grenville Mountains are eroded so deeply that what we walk over today are actually the once deeply buried roots of the mountain belt that have since been brought up to the surface. One can therefore find metamorphic rocks and minerals that only form at the very high temperatures and pressures encountered deep below in the deep crust. And the age of these metamorphic rocks, just over 1 billion years old, turns out to be quite widespread all around the world, suggesting there were many mountain belts of the same age.

If one wanted reconstruct the pre-Pangean supercontinent, then a major first step forward would be figuring out which continent collided with eastern Laurentia. Enter Paul Hoffman. In the early 1990s, before he moved to Harvard University, Hoffman's professional career had been at the Geological Survey of Canada. Even though he would go on to do his most famous work later in the decade, by that time he was already a world-renowned geologist. In his 1991 landmark paper published in the prestigious journal *Science,* Hoffman offered a global reconstruction of Rodinia.[2] In fact, Rodinia was such a new idea at that time, he simply referred to it as the "proposed supercontinent."

Peer review is essential to publishing. If you are reading this book, it is only because it has survived several rounds of review by my fellow geologists. Even though there are certainly creative elements to science, the rules for testing a hypothesis and assessing the outcome of that test must be agreed; otherwise we would just be discussing opinions. Reviewers look at the argument that a proposed publication will make and assess the evidence and methods for collecting that evidence. It surely isn't perfect, as it assumes objectivity in the reviewers, which may not always be the case. The system favors small, incremental change, with the goal of being cautious and correct.

In 1991, Hoffman reviewed two papers for the top topic-specific journal in our field, *Geology,* and in conducting these reviews he ended up forging his own opinions on the nascent notion of Rodinia. Even if a reviewer disagrees with the interpretation of the data the author has provided, if the reviewer cannot find flaw in either the data or the logic of the author, the reviewer should support publication. The two papers Hoffman reviewed by Eldridge Moores and Ian Dalziel, indeed giants in their own right in geology, made significant advances and are still cited today.[3] Although Hoffman could find no flaws in either manuscript and recommended both papers for publication, by his own thinking, they didn't go far enough. On one hand, serving as a reviewer takes time and effort that one otherwise could commit to one's own research; on the other hand, it is an immense privilege to see science before it is published and thus presents an opportunity to the volunteer reviewer to be on the cutting edge if one iterates upon the work soon to be published.

Hoffman recognized that "Grenvillian" mountain belts of the same age as the approximately 1.1 billion-year-old metamorphic rocks from the small village in Quebec could be found on nearly every continent in the world. With one exception, West Africa, this age of mountain building seemed globally significant, and the Grenvillian mountain belts represented the kinds of collisions that might lead to the formation of a supercontinent. Hoffman wasn't the first to note this widespread preponderance of ancient mountains of similar age, but in 1991 he was the first to offer both a proposed configuration of the pre-Pangean supercontinent and a proposed scenario for how it might have transitioned to the modern configuration of continents with which we are familiar.[4] Such a major advance is the kind that gets published in *Science* and typically sets the research agenda of a field of study for years to come.

But the stories behind such great feats always also have a human face. I later learned from one of his past graduate students the backstory of just how Hoffman went about formulating this hypothesis. Although a legendary rock hound, Hoffman has never been at the top of the class in terms of computers. When he was at the Geological Survey of Canada, a cartographer would computer-generate Hoffman's famous geologic maps from his hand-drawn, colored-pencil versions from the field. He is also a firm believer in science as a solitary pursuit. Peer pressure, Hoffman insists, clouds one's thinking when conducting field work, and he was known for typically conducting his day-long geologic traverses alone. Apparently he carried this solitary mentality back to the office. Even though there was computer software available to plot and rotate continents digitally, Hoffman opted instead for cardboard cutouts. Yes, you read that correctly. Cardboard cutouts of the continents.

Paul Hoffman reconstructed supercontinent Rodinia in much the

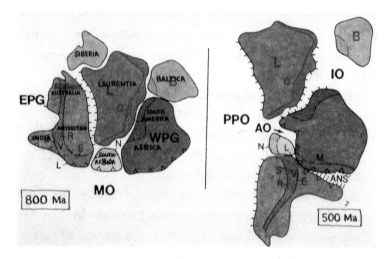

Figure 16. An early version of Hoffman's Rodinia (*left*) and its evolution to Gondwana (*right*) with Laurentia and Baltica near, but separate. Once he had arranged his cardboard cutouts of the continents properly, Hoffman drafted his scientific figure with tracing paper on a light table. Image courtesy of Paul Hoffman.

same way the American Museum of Natural History designs an exercise for children to piece together the continental jigsaw of Pangea (fig. 16). Printing geometrically accurate continental outlines, Hoffman then colored in the Grenvillian mountain belts of each continent and finally got to shuffling the literal puzzle pieces around. He was guided by the shapes of the continental margins (i.e., the indentations and promontories of the puzzle pieces) and the orientation of the Grenvillian mountain ranges (i.e., the part of the picture each puzzle piece depicted). But he was also guided by kinematics: the plate tectonic drift of continents from whatever configuration of Rodinia he preferred had to also transition reasonably well to Pangea. And since the earliest portions of Pangea to assemble were the southern continents which would comprise the megacontinent of Gondwana, it's the position of the Gondwanan continents that he had more clues about.

But although knowing the configuration of the Gondwanan continents in their younger incarnation was certainly a good clue, it remained a very tricky issue to resolve from whence they had come in the Rodinia supercontinent. Compared to the relatively straightforward task of piecing together the rifted fragments of Gondwana in the southern hemisphere, as Alexander Du Toit had done more than a century ago, figuring out the prehistory of the Gondwanan continents is complicated to say the least. Furthermore, it's also tricky business from a practical point of view as it involves doing fieldwork on continents with many developing countries, sometimes politically unstable ones. One of my colleagues in Beijing, Xiaofang He, tells an amazing tale from when she was doing her PhD work in Sri Lanka that involved nothing less than a coup d'état. Xiaofang and her colleagues had completed a successful field season in the beautiful jungle country and had received generous (and delicious!) hospitality by the Sri Lankan citizens throughout their trip. But as she and her team prepared to ship their rock samples and head home, they learned that a military coup had seized control of the country's main airport in Hambantota. Instead of waiting around for bad to go to worse, Xiaofang wisely decided to take evasive action before the situation got more dire. She and her team drove north for hours to another airport in the capital city of Colombo. They escaped just in time. And with the critical rock samples too. Whether dealing with the difficulty of finding outcrops in jungles like Sri Lanka or Madagascar, or even with political unrest, geologists like Xiaofang working on Gondwana provide hard-fought clues to the mysteries of how forming Gondwana involved the reassembly of fragments left over from the breakup of Rodinia. It was these Gondwanan continents that Hoffman had to start rearranging, but he needed another clue as to what continent he might be arranging them around—what was at the heart of Rodinia ?

As about half of Laurentia was bounded by the lengthy Grenville mountain belt along its eastern margin, Hoffman began by surrounding Laurentia with many of the other continents that shared mountain belts of similar Grenvillian age. The Amazonia block of South America, named for the Amazon River that incises it, contains perhaps the best candidate for a mountain belt formed by continental collision with Laurentia. South Africa and Congo are also blocks with Grenvillian mountains that, as Hoffman recognized in 1991, are placed opposite to Laurentia as colliders to form Rodinia.[5] But as is often the case, exceptions to the rule are nearly as insightful as the rule itself. If the rule was that most other continents shared a Grenvillian-aged mountain belt, the exception was West Africa. Of all the continental blocks that would soon comprise Gondwana, West Africa notably doesn't have a Grenvillian orogen. Hoffman thus made a conjecture that still holds to this day, three decades later: those blocks of Gondwana that had Grenvillian orogens collided with Laurentia to form Rodinia in the same mountain building event, whereas West Africa must have been a later addition tacked on to the outside of the supercontinent. These geologic inferences were also consistent with the kinematic transition from Rodinia to Gondwana.

Hoffman realized that the breakup of Rodinia could naturally lead to the assembly of Gondwana. Recall that Laurentia was not a part of Gondwana 500 million years ago, and it would only merge with Europe about 420 million years ago to make Laurussia, which itself would only merge with Gondwana to make Pangea about 350 million years ago. He then put this new clue—Laurentia's later separation from Gondwana—together with the previous clue that much of Laurentia was surrounded by 1.1 billion-year-old orogens. If Laurentia was at the center of Rodinia, then its breakout from the continents around it might have formed Gondwana (fig. 17).

With one simple rotation, Hoffman could transition continents from the breakup of Rodinia to the assembly of Gondwana. The analogy Hoffman used for the sweeping rotation was a Chinese hand fan that pivots about a single hinge. By picking just one point

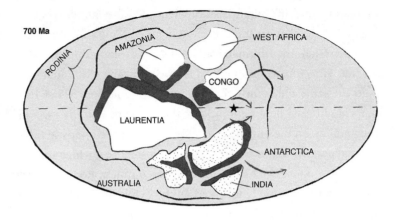

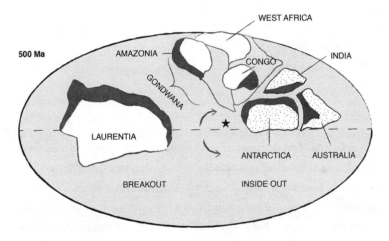

Figure 17. Paul Hoffman's reconstruction of supercontinent Rodinia (700 million years ago [Ma]) and its transition to Gondwana, the first portion of Pangea to form (500 Ma). Hoffman realized that one could take the continents that would later form Gondwana and turn them "inside out" from their positions in his proposed Rodinia in order to transition to the well-known Gondwana. Those continents on either side of Laurentia in Rodinia became the opposite sides of Gondwana: East Gondwana (stippled) and West Gondwana (not stippled). Dark regions were Grenvillian orogens during the assembly of Rodinia. The star marks the point around which Hoffman could rotate the continents on the two sides of Laurentia in Rodinia to create Gondwana. Adapted from P. F. Hoffman, "Did the breakout of Laurentia turn Gondwanaland inside-out?" *Science* 252, no. 5011 (1991): 1409-12.

of rotation south of Laurentia and applying it to all continents breaking away from that central continent in his configuration of supercontinent Rodinia, Hoffman could transition from Rodinia to Gondwana. Thus those continents of later Gondwana with mountain belts the same age as the Grenville—such as Amazonia, which had been landlocked near Laurentia on the inside of Rodinia—would then end up on the outside of Gondwana. And the opposite transitional pattern would be expected for West Africa, which had no Grenville orogen. On the outside of supercontinent Rodinia, West Africa would become landlocked on the inside of Gondwana. By showing that he could create Gondwana by turning the flanks of Laurentia in Rodinia "inside out," Hoffman had made a connection to the younger supercontinent that told us something, for the first time, about the one that had come before it. With these first clues, the hunt for Rodinia was on.

• • •

As sufficient and influential as Hoffman's cardboard cutouts were, Rodinia had to be brought into the twenty-first century. Zheng-Xiang Li would become a central figure in the hunt for Rodinia. Since his landmark paper, published in 2008 with seventeen eminent geologists from around the world backing him up, Li's name has essentially become synonymous with Rodinia—the first supercontinent before Pangea to be taken seriously. He would insist that this was certainly not without a lot of help, but it was his marshaling of diverse opinions from so many international scientists that is perhaps his biggest legacy. Li not only made Rodinia (fig. 18), he forged the way forward for reconstructing all pre-Pangean supercontinents.[6]

Diversity of thinking is perhaps the best way to improve science. Surely this should be the case in *all* areas of science, but never perhaps is the advantage of international collaboration so apparent as when one is trying to discover how the now scattered continents

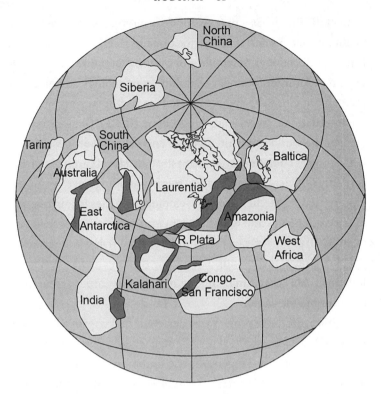

Figure 18. Rodinia according to Li's team. Note many similarities to Hoffman's Rodinia (fig. 17), including the gray shading for Grenvillian orogens. This reconstruction, although similar, was investigated rigorously with paleomagnetism (among other methods). The computer-based reconstruction was generated with GIS software. Adapted from Z. X. Li, S. Bogdanova, A. S. Collins, et al., "Assembly, configuration, and break-up history of Rodinia: A synthesis," *Precambrian Research* 160, no. 1-2 (2008): 179-210.

of the world once fit together in the distant past. If supercontinent Rodinia was truly large like its successor Pangea, then nearly all the continents must have been included. From a very practical perspective, each nation's geologists are most familiar with their regional geology. If one wants to know how each puzzle piece fits into the jigsaw of Rodinia, one is wise to leverage such regional geological knowledge. Li learned this firsthand.

Luckily, there was a preexisting international structure to help facilitate such borderless scientific pursuits: UNESCO, the United Nations Educational, Scientific and Cultural Organization. Providing a hub for international collaboration in geology comes under the remit of UNESCO and is executed via its flagship organization, the International Geoscience Programme (IGCP). Geologists from all over the world can band together and submit a proposal for an IGCP project on their shared interests. IGCP funds aren't huge but certainly sufficient to play a critical role in providing travel for underrepresented researchers from developing countries to join international conferences and workshops. In my time serving as secretary of an IGCP project, we were fortunate to fund rising-star researchers Piotr Krzywiec (Poland), Zhuo Dang (China), and K. V. Wilbert Kehelpannala (Botswana), among many early-career researchers from around the world.

In 1999, an IGCP project was designed and approved to study Rodinia. But even with the green light, it would be an arduous road ahead for Li's team. Tragedy struck, and more than once. This is likely the reason Li hesitates to be referred to as leader of the group, as it was a role he had to step into, and he wasn't alone in having to do so. In truth, there were coleaders, including those that offered support for the group in ways Li could not. Over the course of the several years of the project, three of its coleaders passed away: Raphael Unrug (Rice University), Chris Powell (University of Western Australia), and Henri Kampunzu (University of Botswana).

Like Li, Svetlana Bogdanova of Lund University (Sweden) stepped up as coleader. During the later, difficult years of the project, Bogdanova and Li worked together to motivate members to continue with the ambitious project and to consolidate the project scope. Li says that she played a major role in holding the team together. Bogdanova, a prolific researcher with an infectious personality as a leader, became a leading figure in the assembly of the ancient crust of Scandinavia before passing away a few years ago. One of Bogdan-ova's close colleagues and a coauthor on the 2008 paper, Victoria Pease (Stockholm University), carries on Bogdanova's legacy in her capacity as chief editor of the journal *Precambrian Research*, devoted to studying Earth's oldest rocks.

Supported by the framework of UNESCO-IGCP, Li's team came from all corners of the earth. Nothing less would do if one wanted to reconstruct a supercontinent, as it would take knowing the whole world's geology. It would also take technology to leverage all this knowledge. Around the same time as the plate tectonic revolution was starting to take off in the early 1960s, GIS (Geographic Informa-tion Systems) was also emerging, also thanks to computer advances. GIS is a framework for managing geographic-based data. In the com-puter program, one can create different "layers," each associated with different geographic data sets. This allowed Li's team a huge increase in sophistication and precision from Hoffman's cardboard cutouts. They would be shifting the configuration of the continents around too, but now with computers and vast data sets that were all keyed into precise geographic locations and on a spherical Earth.

Unlike Hoffman's two-tone cardboard cutouts (fig. 16), the conti-nents of Li's team were technicolored. In GIS, they were able to cre-ate color-coded layers for each rock type: igneous, metamorphic, and sedimentary rocks. Then they were able to shade the intensity of each colored layer of different rock types according to geologic age: darker for younger rocks, lighter for older rocks. If two conti-nents had collided to create Rodinia, then one should be able to find green (metamorphic) mountain belts of similar shade (age) on the two continents and these margins of similar color and shade should be facing each other. Next, younger igneous and sedimentary rocks could be used to test this metamorphic-based configuration. If indeed the two continents had been nearest neighbors in Rodinia, then when the supercontinent broke up, there should have also been geologic evidence of it. The stretching of two continents that eventually leads to their later separation gives rise to both sedimentary basins deposited in the valleys of drawn-down blocks of crust as well as igneous rocks intruding up from the hot mantle causing continental breakup, both of which are happening in the East African Rift valley today. Then, as oceans open up between the separating continents, continental margins form shelves of sediments that are excellently preserved in the geologic record. In this way, Li's team was taking full advantage of both precise geographic computer-based global data sets as well as the entirety of the diverse geologic record. No evidence would go to waste.

The geologic correlation of different rock types in Li's color-coded GIS layers boils down to one question: How old is a rock? Ultimately a leading question, the answer is required to solve most problems in geology, and solving the geography of ancient supercontinents is no exception. There is no one way to date a rock, and indeed different methods are more useful depending on whether the rock is most likely young or potentially very old. By far the most commonly used method of "geochronology" is radiometric dating, which uses the systematic, long-term decay of radioactive elements over time to back-calculate when a rock formed. The radioactive decay of uranium to lead as measured in the resilient mineral zircon is widely considered the gold standard of deep-time geochronology.

Fortunately for geochronologists, the disintegration of uranium happens slowly, very slowly. A half-life of a radioactive substance is the amount of time it takes for half a sample to decay. The half-life of uranium-238 (the flavor, or isotope, we use), is actually similar to the age of the earth, 4.5 billion years. This means that there is plenty of uranium left to measure, but that we also need very sensitive instruments in order to detect what incredibly small quantity has decayed into lead, and isotopes of lead along the way to get there.

•    •    •

Another critical technique that Li's team used was our old friend paleomagnetism. We discussed paleomagnetism earlier as it was one of the emerging scientific tools that helped test the existence of Pangea. Recall Ted Irving's underappreciated, but later vindicated, Cambridge thesis work. As plate tectonics itself, in addition to Pangea, was then being put on trial, any available test was crucial to conduct at that time. The case for Pangea had been greatly aided by the discovery of seafloor spreading, which provided a mechanism to push the continents apart. It was the seafloor record that convinced most people of the continental drift that broke apart Pangea. However, no such seafloor record exists to guide us with Rodinia: most, if not nearly all, the ocean crust of that antiquity has long since been subducted and recycled back into the mantle. Whereas paleomagnetism had largely played a supporting role as corroborating evidence for Pangea, when it came to testing Rodinia, it would move into a leading role.

After all the hard work of sampling the rocks and carefully measuring their magnetic directions in the laboratory, a paleomagnetic study boils down to two numbers: the latitude and longitude of a paleomagnetic pole. My PhD advisor at Yale, David Evans, teaches that paleomagnetic poles can be thought of as ambassadors of the continents they come from. Move the continent, and one must move the paleomagnetic pole along with it. Poles and continents are rigidly connected, much like ambassadors must have the interests of the countries they represent in mind even though they live abroad. Although its definition may sound a tad abstract, a paleomagnetic pole is the location of the magnetic pole relative to a continent. If we assume that the magnetic poles have always been in the same place (over long timescales), then if a paleomagnetic pole points anywhere but the North (or South) Pole, then we must conclude that the continent has moved; the location of its paleomagnetic pole tells us how and how far.

In order to reconstruct the ancient position of the continent at the

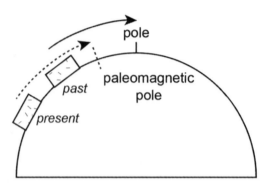

Figure 19. How paleomagnetic poles are used to reconstruct ancient continental positions. The measurement of inclination points to the location of the paleomagnetic pole (dashed lines). The paleomagnetic pole is shifted to the North Pole and the continent is shifted along with it (solid lines). The past position of the continent was closer to the pole (higher paleolatitude) than the continent's present position, as in the South Dakota example.

age that the rock formed, one simply rotates the paleomagnetic pole to the North Pole to get its original location. The continent thus gets dragged as far as it must in order to bring its paleomagnetic pole to the North Pole (fig. 19). For example, if we sampled rock from South Dakota and the paleomagnetic pole we measured plots in northern Canada, this would imply that South Dakota was at a higher latitude at the time the rock formed because we have to shift the paleopole to the present-day North Pole and bring the continent along with it in order to reconstruct the ancient latitude of South Dakota.

Compiling paleomagnetic poles from all the continents, Li and his team were not just matching rocks with like rocks. They were also making sure that the continents were at their appropriate latitudes at the appropriate ages.

Unfortunately the reliability of a paleomagnetic pole can be called into question in many ways—and we will only mention the most obvious pitfalls. Even if a rock is stably magnetized when it forms, it can later be remagnetized, particularly if the rock is quite old and metamorphosed. Imagine a lava, erupted 1 billion years ago at the time of Rodinia, having cooled and acquired a stable magnetization. This original information still has to survive unaffected for 1 billion years in order for it to be useful to us today. Also imagine if the mountain building associated with the assembly of the subsequent supercontinent Pangea caused these once-fresh lava flows to be subjected to great heat for long periods of time. If they were heated enough, or for long enough, the magnetic domains in the magnetic minerals in the lava flow (such as magnetite) would be given enough energy to be realigned with the orientation of the magnetic field at that younger age of Pangea, thus losing their original orientation from the time of Rodinia forever.

Other potential issues remain even if the rock avoids remagnetization. Just because a rock is stably magnetized and yields wonderful paleomagnetic data when measured on a magnetometer doesn't mean the same rock can also be reliably dated to know its age. That is, even if we think the magnetization was acquired when the rock formed (it is the primary magnetization), if the age of the rock cannot be accurately determined, then the calculated paleomagnetic pole, however reliable, is essentially useless. This is certainly a case of the scientific crossword puzzle at work, where both reliable paleomagnetism and geochronology are needed to solve the puzzle. Both types of experts are required in reconstructing paleogeography. As we will discover later, completing this particular crossword puzzle is not always made easy due to the diverse compositions of rocks. With all these complications in mind, Li's team waded through the wealth of data for the true riches that could be used to reconstruct Rodinia.

•    •    •

After rigorous data filtering, scientists hope there's enough data left to say something meaningful. This concern is because whether you have only one paleomagnetic pole from a continent, or multiple poles of different ages, critically influences the precision with which a continent can be configured in a supercontinent. Of course having just one pole, however disappointing, is certainly better than nothing. With that one pole, a paleomagnetist is able to rotate the pole (and the continent along with) to the North Pole (fig. 19). In this orientation and at this latitude, then the paleomagnetist can swivel the continent around the globe, testing all possible longitudinal positions without changing the continent's latitude or orientation. This is why paleomagnetists often say that their data constrain latitude but not longitude. Earth's dipolar magnetic field is identical at all longitudes. There is no difference in the orientation of magnetic minerals in rocks formed at the same latitude, even if they are thousands of kilometers apart in longitude. When only one pole is available for a continent within a given age range, a continent's longitude is unknown. This ambiguity requires the geologist to consider several different potential configurations relative to different continents at the same latitude, and they must rely more on geologic correlations to test between such options.

But everything changes if two or more paleomagnetic poles can be obtained for rocks of different ages in the same continent, and you can be much more certain about the position of the continent in an ancient supercontinent. If two, three, or even more poles from a continent through a certain time interval of Earth history can be obtained, one can trace a path between the poles. Paleomagnetists call this an "apparent polar wander" path. This term is used because although the path of the pole apparently depicts the North Pole moving over time, it is actually the continent drifting due to plate tectonics relative to the stationary North Pole that gives rise to this wandering path. If two continents were connected in a supercontinent, then they should share the same wandering path. In addition, their paths of motion over time should: (1) converge when they collided, (2) be the same during the lifetime of the supercontinent, and (3) then diverge once the two continents breakaway from each other (fig. 20).

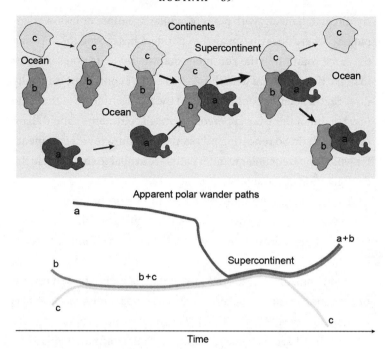

Figure 20. Illustration of how continents connected in a supercontinent should share the same "apparent polar wander" paths constructed from paleomagnetic poles for the time the supercontinent existed. During assembly, the paths converge; during breakup, the paths diverge. The paths are generated by connecting the dots between measured paleomagnetic poles. Adapted from D. A. D. Evans and S. A. Pisarevsky, "Plate tectonics on early Earth? Weighing the paleomagnetic evidence," *When did plate tectonics begin on planet Earth?* 440 (2008): 249–63, and R. N. Mitchell, N. Zhang, J. Salminen, et al., "The supercontinent cycle," *Nature Reviews Earth & Environment* 2, no. 5 (2021): 358–74.

But the benefit of such a shared path of motion goes beyond telling you whether two continents were connected and for how long—it also tells you *how* they were connected. Treating the poles again like ambassadors, connected to their continents, we rotate the poles of one continent (and the continent they are connected to) so that the poles are superimposed on those of the other continent. Since the polar paths are similar in shape to each other, superimposing them should match nicely, with poles of similar age more or less overlapping. So where does this mean the continent has moved to? Applying the same rotation to the continent that we did to its poles in order to make them overlap with those of the other continent, we finally discover the relative configuration of the two continents at that time (fig. 20). But if there is a supercontinent, all continents should share the same path. So repeating this same procedure for all continents for which apparent polar wander paths are available should yield the configuration of the supercontinent.

If a supercontinent indeed existed, then most of the continents of the world should share a similar polar path during that time. This was one of the more definitive tests that Li's team attempted to conduct for Rodinia. Unfortunately, not all continents' data sets are created equal, and the availability of data is highly biased toward richer countries like the United States, Canada, and Australia. This is where the role of UNESCO in IGCP projects becomes so critical to the future of Earth science research. As Paul Hoffman liked to say, "If one wants to reconstruct a supercontinent, then one simply has to learn the world's geology." While this is indeed the long-term goal, Li's team had to work with what they had at that time. While continents like Laurentia had a wealth of paleomagnetic poles from North American countries, less-studied continental blocks like West Africa had few or none. So, just the way you would go about completing a jigsaw puzzle, you obviously start with the easier pieces to place and leave the difficult, nondescript pieces for the end. Once the puzzle is almost solved, then the remaining possibilities to place these trickier "blue sky" pieces are greatly reduced.

As luck would have it, Laurentia was a good piece of the puzzle to have a lot of data from. Laurentia, it turned out, was likely at the center of the Rodinia supercontinent. Then the other continental blocks could be positioned based on paleomagnetism or geologic correlations (but ideally both) around Laurentia along its different continental margins. Australia also had a lot of paleomagnetic data like Laurentia, but even with this wealth of data, positioning these two continents relative to each other proved a lot harder than one would have expected. Because of this, Li's team weighed the pros and cons of as many as *four* different relative configurations of Australia and Laurentia—and the one they ultimately chose didn't even have a direct connection between them, but had South China positioned in between them as a missing link.[7] Only several years later, in 2011, would Li and my PhD advisor, David Evans, discover why, despite so much data, configuring these two continents had proven so surprisingly difficult.

Li and Evans eventually realized that there was a major complication with the data from Australia. They noticed that paleomagnetic poles collected from rocks in southern Australia didn't agree with those collected from northern Australia. Continents are largely rigid blocks, but even the oldest continents like Australia and Laurentia are composed of multiple smaller blocks. Luckily, the Australian record had been sampled in such detail that several ages of rocks had been sampled from both the southern and northern halves of Australia. It turned out that the results from the two halves of the outback didn't agree as they should have. However, the mismatches were systematic: none of poles of the same age matched, but the mismatches were approximately all the same size and in the same direction. Li and Evans thus realized that if they rotated the poles of one of the blocks, that single rotation could make the poles of the two halves at the various ages come into agreement.[8] They had discovered an ancient tectonic rotation *within* the Australian plate itself. This intraplate rotation of Australia had occurred when India slammed into Australia during the assembly of Gondwana, causing the two halves of Australia to shift relative to each other. Since this Gondwana-aged shift within Australia was younger than Rodinia, one had to consider the original configuration of north and south Australia in order to make a coherent apparent polar wander path with which to compare Laurentia. Positioning Australia not far from the western margin of Laurentia was now done much more confidently than ever before. Li and Evans titled their paper, "A Tighter-Fitting, Longer-Lasting Rodinia." Have I mentioned that geologists typically don't take themselves too seriously?

But this minor readjustment of the Australian data was the least of the challenges facing Li's team. And unfortunately the predicament facing them then still largely stands today. One of the most critical pieces of the putative Rodinia puzzle essentially had no paleomagnetic data to speak of: Amazonia, the continental block underlying the Amazon River in northern South America. Fortunately, as Li's interdisciplinary team emphasized, there's not just one way to reconstruct a supercontinent, and what Amazonia lacked (and still lacks) in paleomagnetic poles, it made up for in geologic evidence. Amazonia had evidence of a mountain belt, the Sunsas orogen, that was the same age as the all-important Grenville orogen of Laurentia. In this respect, Hoffman's cardboard cutouts had been sufficient: his placement of Amazonia was nearly identical to Li's sophisticated GIS-based Rodinia reconstruction.

Amazonia is the best candidate for the continent that collided with Laurentia to form the central spine of the Grenville Mountains. Why do we think this? Recall that from as early as Hoffman's Rodinia work, a handful of continents were placed along the spine of the Grenville orogen. But Amazonia, then and now, is placed most centrally opposite the Grenville front, and with most confidence. The smoking gun comes in the comparison of the lead isotopes of rocks in Amazonia and Laurentia. In addition to the metamorphic rocks that form under the immense pressure and temperature that accompany continental collisions, many igneous rocks, namely granites, also form—a lot of them. Before we see how the isotopes of a certain granite can serve like a fingerprint to identify it, it's worth discussing why granite itself is so important.

•   •   •

Along with plate tectonics, complex life, and an ocean, granite is arguably one of the defining characteristics of Earth. We've already discussed how the buoyancy of granite allows the continents to ride high relative to the dense, basaltic oceanic crust. The Moon, Mars, Venus, and Mercury all have basalt. But none of those other rocky bodies have granite. Sure, the Moon and Mars have some more "evolved" igneous rocks that have started leaning in a direction away from basalt and toward granite, but no card-carrying igneous petrologist (experts in igneous rocks) would dare to classify these intermediate compositions as granite sensu stricto. Igneous rock types form an evolutionary chain, starting with "primitive" forms like basalt and ending with "evolved" forms like granite (this is not my overuse of metaphor, these are actually the terms petrologists use). So how do primitive rocks create more evolved rocks? Let's quickly discuss the two main ways these all-important and complex rocks come into being.

Believe it or not, many magma chambers may start with primitive compositions but naturally evolve toward creating granite magma as they cool. Different minerals crystallize at different temperatures: some like it hot, others like it cold. Because of this, the composition of a magma actually changes as it cools. Those minerals that comprise a basalt (olivine, pyroxene, and calcium-rich plagioclase) all crystallize at high temperatures, and any magma escaping from that chamber to fuel a volcano above it tends to be basaltic in composition. Basalt consists of about 55% silicon dioxide ($SiO_2$), also known as silica; granite about 75% $SiO_2$. So when the magmas crystallize, basalt contains much less of the mineral quartz (whose chemical composition is $SiO_2$) than granite. The main compositional difference between basalt and granite is $SiO_2$ content: basalt has very little to no quartz and granite is defined as having between 20% and as much as 60% of quartz by volume. This evolution in composition is facilitated by a process called fractional crystallization. The composition of the melt changes as the early-forming minerals crystallize at high temperature. These minerals are also dense and so sink to the bottom of the chamber and are essentially removed from the remaining (or residual) melt. Thus, the mere act of crystallizing minerals (removing them from the residual melt) changes the magma chemistry to more "evolved" compositions so that when it crystallizes, the rock formed is granite, dominated by minerals such as mica, sodium-rich plagioclase, and of course, quartz. But it turns out there's an even easier way to form granite: just add water!

Another way to make granite, and possibly more widely applicable to forming the bulk of Earth's crust, is simply by melting basalt. And it turns out the role of water in melting is absolutely critical. Water melts basalt by dissolving it. Just like the food-encrusted fry pan you left to soak overnight, water dissolves crusty rock too. Water is known as the universal solvent because it dissolves more substances than any other liquid. This is due to its polarized molecular structure, with positively charged hydrogen on one side and negatively charged oxygen on the other. With this ability to become attracted to many different types of molecules, the polarized water molecule is able to overcome the bonding forces of many molecules by tugging at and weakening their bonds, forcing salt (NaCl), for example, to dissolve and breakdown into separate sodium and chlorine ions. This seemingly magic liquid also dissolves rock deep in Earth's crust.

As long as water is present deep in the crust, basalt will melt at low temperatures, and this melt will form granite.[9] But how does the water get deep in the crust in the first place? Our favorite one-word answer: subduction. As seawater is liberated from the sinking oceanic plate, it rises and gets concentrated over time; eventually it percolates into the deep crust of the upper plate above it, which is similar to basalt in composition. It is this deep crust at the base of the upper plate that is melted to create granite. When subduction occurs in one place for a long time, not only do a lot of granites form, but it also becomes more inevitable that a continental collision becomes imminent. With so much oceanic crust consumed, a continent can't be too far behind. Therefore, the blooming of granites is the prelude to, or a warning sign of, the impending collision of continents.

It is the granite bloom in the Grenville orogen that tells us which continents collided with which in the formation of Rodinia. Geochronology, as we have discussed, is our first step in following the evidence. For magma chambers of colliding continents to have been related, then they must have similar-aged granites. But matching ages alone isn't proof enough. Following our maxim "the present is the key to the past," granites are forming today in both the Andes and the Himalayas, but these contemporary orogens are on opposite ends of the earth. So how does granite, important as it is on Earth, exactly help us solve Rodinia? Again we reply on isotopes, this time not of uranium but of lead.

•   •   •

Radioactivity is important not only for precisely dating the age of a rock but also for fingerprinting the composition of the source of the magma from which the granite formed. As we've just seen, granites are formed by melting the crust beneath them, which is called the "basement" by geologists. If two continents had indeed collided, then the magmas generated at that time should share similar sources of melting, those sources being the basements of the two continents in question. Recall that uranium isotopes are constantly and slowly decaying to lead. If two granites came from the same magma source, they should not only be the same age, they should have also evolved together. We have a test: two granites from the same source of magma should have the same original uranium-to-lead ratio (i.e., at the time of crystallization), as well as the same ratios of lead isotopes produced since that time as the uranium decayed to lead.

Now, for the most part, most of the granites from Laurentia and Amazonia have different levels of lead isotopes. This might at first appear to contradict the potential correlation between the Grenville orogen of Laurentia and the Sunsas orogen of Amazonia. But not all hope is lost. Recall that granites are made by melting the crust. And that all genuinely new crust was originally extracted from the mantle. So depending on whether that crust is young or old—that is, if it was extracted from the mantle recently or a while ago—it will have low or high initial levels of lead isotopes, respectively. As the basements of Laurentia and Amazonia were extracted from their respective mantles at different times, when their respective crusts are melted to form granites, their differing lead isotopes naturally betray their different ancient crustal heritage. In this sense, lead isotopes are very useful for detecting that an apparently coherent continent is actually comprised of two once-very-different and separated pieces of crust that have been tectonically juxtaposed by continent collision. So how exactly do lead isotopes help us test the Laurentia-Amazonia connection in Rodinia?

Fortunately, we have more clues. Although we have spoken of the Grenville orogen as a large, single mountain belt formed by colliding continents, in truth it also contains a complex assemblage of microcontinents—often called "terranes"—that got smashed together. If you look at a map of the West Pacific, it is dotted with a multitude of islands. Should the West Pacific be involved in a continent-continent collision in the future, those intervening islands would be dismembered, but their crust would be trapped in the vice-grip of the converging continents and preserved as a complex mosaic of separate terranes. Terranes participate in plate tectonics just like the major continents, and each has its distinct geological history and structure. While terranes may have short-lived periods of tectonic independence, they get swept into collisions between large continents and lose their independence, possibly forever.

Terranes can provide critical missing links between two continents that are suspected to have collided. Terranes follow the same tectonic rules as the larger continents around them, and they become embedded in larger continents after their independent streak has run its course. It turns out the Grenville orogen contains a diverse collage of both mountains of granite (called "massifs") as well as terranes. The edges of continents get worked and reworked due to repeated tectonic cycles from one supercontinent to the next. Although the Appalachian Mountains are most famous for their role in the assembly of Pangea, ancestral Appalachia was also involved in the assembly of Rodinia. As such, remnants of the Grenville orogen are in fact found in the basement rocks beneath the Paleozoic rocks of the southern Appalachians.

It is in the southern Appalachians, such as the Blue Ridge Mountains of Tennessee and the Carolinas, that we find the clue that links Laurentia (of North America) and Amazonia (of South America) together at the heart of Rodinia. Terranes that collide with continental margins are a prequel for the main event: the eventual collision of continents that flank the narrowing ocean. In the southern Appalachians, at the edge of the Grenville orogen, there are "exotic" terranes, meaning that they came from far away. We know this because the lead isotopes of the granites of the terranes are completely different from those of the granites of the rest of the Grenville orogen. The granites in the southern Appalachians, although the same age as the Grenville orogen, could not have been sourced from the same magma as those that formed by melting Laurentian crust. A better match for these southern Appalachian rocks might be found in Amazonia.

Although she had the vision to venture to South America to test the possible link between Amazonia and Appalachia, it was Erin Martin's first time conducting field work as a graduate student, and she learned quickly that nothing in the field goes according to plan. It turned out she would have fewer problems with the complex rock formations than she did navigating the dialect of Spanish used in that part of Argentina. (For example, the pronunciation of "tortilla" doesn't have the "y" sound for the double "l" but sounds instead like "ch": "torticha.") Imagine trying to direct a tow truck driver to your field vehicle stranded somewhere in a remote landscape when it's difficult to articulate even the location of the nearest town. But Martin persisted, somehow getting her truck fixed and on the road in less than a day. But her troubles had neither started nor stopped there. She had assumed she could find basic field equipment upon arrival. After finding no large hardware store, she scoured multiple local small hardware stores for a large sledgehammer (critical for sampling granites rich in the hard mineral quartz). The one she bought immediately split at the handle upon the very first strike of the very first rock. But Martin persisted. She bought a trusty old sledge from a farmer. It lasted her the whole trip. Committed to learning her lesson of always arriving prepared, she brought it home with her to Australia, pledging to bring her trusty sledge with her wherever she travels for fieldwork next. Impressively overcoming each and every setback, Martin collected all the samples she needed from Amazonia and would eventually go on to successfully earn her PhD.

Previous studies had shown that Amazonia has a lead isotope fingerprint that was distinct from that of Laurentia. The question thus became, if Appalachia wasn't from Laurentia, could Martin prove it was from Amazonia? When Martin compared them, she found that the lead isotopes of the southern Appalachian terrane were indeed identical to those of Amazonia.[10] The southern Appalachians now embedded in the Grenville orogen of Laurentia had come from Amazonia. That is, the granites of the southern Appalachians had formed by the melting of the crustal basement of Amazonia. Then the rogue terrane broke away from Amazonia and lived out a short time of tectonic independence, during which it crossed an entire ocean on its own, no longer part of Amazonia but not yet part of Laurentia. Eventually the terrane collided with Laurentia, where it got stacked up alongside the granite massifs of Laurentia with similar ages but with radically different lead isotopes. Then, like the terrane it had sent as an emissary in advance, Amazonia followed suit, ultimately also colliding with Laurentia. In this final phase of collision, the Grenville Mountains were thrust high up in the sky as the two major continents collided and supercontinent Rodinia was finally formed.

Figure 18 shows the final reconstruction of Rodinia that Li's group created. The team of geologists from all over the world had literally put Rodinia on the map. Undoubtedly some adjustments have been made and will continue to be made in years to come. But many of the most fundamental connections, like Laurentia and Amazonia across the Grenville orogen, are likely to stand the test of time. And there are also other important things we have learned about Rodinia aside from what the supercontinent simply looked like. We have also learned how Rodinia behaved and how it shaped climate and life on Earth. Once a supercontinent forms, it changes how the planet operates by changing the shape of the earth, the stability of Earth's rotation, and even the climate and the environmental conditions for the evolution of life.

Rodinia is perhaps most known for the deep freeze it experienced: the most severe glaciation in all of Earth history. Known as "Snowball Earth," just as it sounds, the entire globe was encapsulated in ice (including its oceans). In the modern world, Earth just has polar ice caps (think Antarctica and Greenland) where incoming solar radiation is less, due to the high angle at which photons arrive at high latitudes. But imagine if polar ice caps grew larger and larger—eventually, if they became large enough, the increasingly white planet would eventually start reflecting more light than it absorbed. As any skier will attest, snow and ice are highly reflective. At such a tipping point, the whole Earth would rapidly become enveloped in ice, from the poles to the equator, and this is thought to have happened twice during the time of Rodinia. The verdict is still out as to why it happened to Rodinia but not to the other two supercontinents. But one thing is clear: by the time Rodinia started breaking up, most of its continents straddled the equator. It turns out this is a recipe for a climate disaster, because putting all the continents in the tropical weathering belt takes a lot of carbon dioxide out of the atmosphere during chemical weathering of rocks. As opposed to rising $CO_2$ levels that trigger global warming, significantly lowered $CO_2$ can cause catastrophic global cooling. But the stress of a global snowball apparently wasn't all bad; on the contrary, the ecological natural selection that its extreme conditions imposed made it such that those life-forms that did survive were poised and ready on their marks after the deep freeze melted. The rifting of Rodinia was thus home to the first algae and multicellular life (Metazoa) that would someday soon become large animals.

·   ·   ·

So far we have established the reality of Rodinia, which is certainly important for the very existence of a supercontinent cycle. But we have yet to explore why Rodinia is really important in our story—that is, what clues Rodinia may offer us about predicting the next supercontinent. And this is where we pick up again my tale from the Australian outback. We left off at the campfire where Professors Maloof and Halverson were telling us stories of when they were young students of the demanding Hoffman.

Our week or so of geotourism through the stunning Flinders Ranges National Park of South Australia had just come to a close. The Flinders Ranges are truly breathtaking geologically. This area of South Australia had once been a deep rift in the continent that filled up with sedimentary layers. Then, when tectonic forces changed from extension to compression, the rift ended and the strata got folded into the spectacular landforms we see now. The Flinders strata contain some of the world's best-preserved large-scale multicellular life-forms, known as the Ediacaran fauna. These peculiar life-forms have even been likened to aliens, they so little resemble life today. The famous *Dickinsonia*, for example, looks more like a fern than any animal (fig. 21). Most importantly, they are the oldest

Figure 21. Ediacaran fossil, *Dickinsonia*, preserving no hard parts but only the imprint of the soft-bodied animal (or plant?) in a sandstone of the Flinders Ranges, South Australia. Specimen shown from the South Australia Museum, Adelaide. Photo credit: Dr. Alex Liu, University of Cambridge.

known animals that are visible without the help of a microscope, as large as 4 feet (~1.5 m) or so in length but only a few millimeters thick, like massive flapjacks. But these first large animals to appear in Ediacaran time were soft-bodied and went extinct before the following Cambrian period. How the Ediacaran metazoans relate, or not, to

Figure 22. The field area at Ross River (running down the middle of the photo). Dramatic tilting, folding, and faulting of the sedimentary rock layers is apparent. Individual stratigraphic layers can be followed for large distances. Image is about 37 miles (~60 km) wide. Courtesy of the public domain from https://earthexplorer.usgs.gov.

the beginning of more familiar life in the Cambrian explosion millions of years later remains a mystery. But the sightseeing was over, and it was time for Swanson-Hysell and me to get back to work.

With the professors having headed back to their respective universities, Swanson-Hysell and I were on our own. It just being the two of us, we got into a very regular routine. We quickly got Swanson-Hysell the critical samples he would need to analyze back in the Princeton laboratory for his PhD thesis. After several successful weeks of sampling, we had filled many sample bags and were in the latter half of the trip—of course, isn't this always when tragedy seems to strike, with the end in sight?

We were in the MacDonnell Ranges of Central Australia camped at a place called—what seemed at first a fitting name for a place of my own—Ross River (fig. 22). Our camp was simple: two tents, a table, and two chairs around the campfire. We did most our cooking over the fire, using our gas stove only for coffee and pasta. Our truck was the Australian outback standard-bearer, a Toyota Hilux, with a canopy over the pickup that served two purposes. In the field, it kept the ubiquitous dust out and the harsh sun off our equipment; and when we resupplied in town, it kept our rather expensive (and hard to replace) field gear from being stolen. Other than our rock hammers and geological field gear, we didn't have much else—just a "footy" (an Australian rules football) to kick around for fun, exercise, and local indoctrination, and a shovel for . . . well, for when nature called. And we might have also had a Frisbee that ended up serving more critically as a second cutting board to expedite the cooking for two hungry blokes. It was a simple life, and that's the way we liked it. Just us and the rocks.

Field work in the Australian outback is particularly special because one can literally camp on the outcrop. From the comfort of our tents, we could see the very sedimentary strata we would sample that day (fig. 23). We had already spent a few days working the Ross River section. A pile of sedimentary strata that is well exposed is referred to by geologists as a stratigraphic "section." First, we measured the section. This involves taking a retractable ruler, holding it perpendicular to the strata, and measuring the thickness of the sedimentary layers. Because tectonic forces have taken layers that were once deposited horizontally and tilted them, the layers of the Ross River section were dipping at about 45° (the dip angle in the picture seems slightly shallower as the layers are dipping slightly into the hill as well as off the left). Such a steep dip angle conveniently allows one to walk from layer to layer without having to walk very far—thanks tectonics! Now, we didn't measure the thickness of every single sedimentary layer, which would have taken forever. What we did was measure the thickness of layers of the same composition, be it sand, shale, limestone or what have you, and then we went on to the next package of similar sediment type.

What we mostly had on our hands at Ross River were carbonate sedimentary rocks. Limestone, the best-known carbonate rock,

Figure 23. Our campsite at Ross River. My tent and drilling water pump are in the foreground; Nicholas Swanson-Hysell, our Toyota Hilux, campfire, and camp kitchen in front of the sedimentary strata in the background. Photo credit: Ross Mitchell.

is made up of calcium carbonate ($CaCO_3$). But in our section, we had a lot of dolomite, which is a magnesium-rich carbonate rock ($CaMgCO_3$). For reasons that are still largely unknown, dolomite is much more commonly found than limestone in ancient rocks, and the opposite is true of more modern strata, where limestone now predominates over dolomite. As we were working during the antiquated time of Rodinia about 800 million years ago, we were therefore dealing with a lot of dolomite.

Aside from their dolomitic composition, the rocks we were sampling were most intriguing because of their shape and form. We were dealing with a very special variety of carbonate rock formed from fossils called "stromatolites." These microbial mounds are perhaps our earliest signs of life on Earth—not the particular ones we were sampling at Ross River, but there are stromatolites more than 3 billion

Figure 24. Stromatolites group together to form an ancient organic mound-like reef built by cyanobacteria, as seen in these parallel, vertically stacked mounds. This sample from China, over 1 billion years old, is about 13 centimeters tall. Photo credit: Ross Mitchell.

years old that argue for at least bacterial life being present on Earth very early in its evolution.[11] Stromatolites are chalky mounds constructed with sediment trapped by lime-secreting cyanobacteria—the type of bacteria capable of photosynthesis. Stromatolites are relatively common in Precambrian rocks and are the earliest form of life on Earth (fig. 24). Since cyanobacteria are still alive today, one can travel to the remote Shark Bay of western Australia to see how stromatolites actually form. Depending on the type of microbial community, stromatolites can come in the form of narrow arching stacks on the scale of decimeters, or wide, broad "bioherms" on the scale of meters. In addition to the mesmerizing structures they left behind, cyanobacteria, as organisms that photosynthesize, have arguably an even more profound legacy: they produced the oxygen we breathe in the atmosphere. But that part of the story will have to wait.

Since we were dealing with carbonate rocks, Swanson-Hysell was able to include a fascinating area of study in his thesis: the evolution of ocean chemistry during the time of Rodinia. The carbon content in the stromatolites could tell us what the composition of the oceans was like then and if and how it changed. You likely already know about the water cycle: liquid water evaporates into water vapor, condenses to form clouds, then precipitates as rain to start the cycle all over again. There is also a carbon cycle, and you may very well know about this due to the current climate crisis we are facing today. Like water, carbon is exchanged around Earth's different spheres: the atmosphere, the biosphere, the ocean, and even the rocks or "lithosphere." Carbon is a major component of limestone and dolomite, and these sedimentary rocks are a critical reservoir of carbon in its journey around the Earth system. So what does the carbon cycle tell us about the oceans and what did Swanson-Hysell expect to find?

Isotopes, once again, come into play. We've discussed uranium and lead isotopes, which radioactively decay from one isotope to another along a series from uranium to lead. Carbon has isotopes of its own, most famously carbon-14, carbon-13, and carbon-12, with the number indicating the number of subatomic particles (neutrons) in the atom's nucleus. Like the uranium-to-lead series isotopes, carbon-14 is a "radioisotope" that decays radioactively over time. Whereas the half-life of uranium is as long as the age of the earth, making it useful to date very old rocks, the half-life of carbon-14 is, geologically speaking at least, very short at 5,730 years, which means it decays very quickly. In rocks that are millions of years old, there is so little carbon-14 left that it is not practical to measure it. Because of this, you will likely be familiar with carbon dating (the decay of radioactive carbon-14), which is used to date archaeological artifacts. But the isotopes of interest to us now, carbon-13 and carbon-12, are stable isotopes that are not radioactive and do not spontaneously decay. They are useful to us for another reason: because of how biology interacts with them. The difference between the stable isotopes of carbon-12 and carbon-13 is a matter of mass and this slight difference can again, as with radioactive isotopes, be measured by scientists. But this slight difference in mass makes a big difference when it comes to photosynthesis. Plants and organisms prefer the light isotope (carbon-12) when converting sunlight and $CO_2$ into energy (sugars) to fuel their activities. There might not have been plants yet during the age of Rodinia, but the supercontinent certainly had cyanobacteria as evidenced in the stromatolites Swanson-Hysell and I were sampling. During times when cyanobacteria and other photosynthesizing organisms were abundant in the surface waters, they gobbled up the lighter isotope as it's just easier for them to absorb. Volcanoes are constantly spewing carbon dioxide into the atmosphere: but only 1% is carbon-13, with the rest being carbon-12. Therefore, only life can concentrate carbon-13 in the oceans, and it does so by preferentially absorbing lots of the carbon-12 from volcanoes. But without algae and bacteria to use up the lighter carbon-12 to make their soft tissues, that is, with volcanoes left to their own devices, carbon-13 does not get concentrated in the oceans relative to the more abundant carbon-12. In short, a lack of carbon-13 suggests a lack of life.

And that's exactly what Swanson-Hysell found in the oceans at the time of the strata exposed at Ross River: they plunged down to volcanic carbon-12-rich values, showing little evidence for the effects of biology.[12] For whatever reason, the environmental conditions at that time became globally unfavorable for fostering life. Without life teeming in the oceans, few microscopic fossils and little organic material formed to get buried in the rock record. What was also surprising was how quickly this lifeless world came and went. The sedimentary layers both above and below the layers at Ross River were completely normal, with background levels of carbon-13 that suggested thriving biologic activity in the oceans; whereas the stratigraphic section at Ross River was another world altogether and, in geological terms, the apocalypse had happened overnight. But in addition to the quickness of the environmental changes, it was also striking how temporary the disturbance was. Although the oceans became relatively lifeless for a few million years at the time the Ross River strata were deposited, afterward essentially the same biologically teeming oceans returned. Something caused life in the oceans to suddenly die off; but then as quickly as the plague had come, it also went.

Furthermore, it didn't just look like disaster was followed by a recovery. For comparison, when the asteroid that killed the dinosaurs (and many other animals) hit some 66 million years ago, the oceans similarly plunged to volcanic carbon values as the impact killed off most photosynthetic organisms living in the surface ocean. Life eventually came back many thousands of years later, allowing the carbon cycle to return to carbon-13-rich values as organisms started to use more and more carbon-12 again. Like a stock market crash, the asteroid impact was a sudden apocalypse followed by a slow recovery. But this pattern was different from the pattern at Ross River. Both the dying off and the recovery at Ross River were equally as sudden. It was almost as if whatever had caused the dying off, oddly enough, had reversed itself. Another difference from the asteroid impact was that although the changes at Ross River were geologically rapid, they were by no means as instantaneous as an asteroid suddenly hitting Earth one day. While the environmental changes at Ross River were surprisingly abrupt geologically speaking, they were in fact portended in the thousands of years of sedimentary layers laid down before the total shutdown. Yet as much as there might have been a brief warning of the biologically razed world to come, the sudden change at Ross River occurred faster than most geologic processes we typically think of—and almost certainly too fast for the fingernail-growing pace of plate tectonics to have played a role.

•   •   •

The reason Swanson-Hysell was conducting his PhD fieldwork in the Australian outback was because he and his advisor, Maloof, had a hypothesis, and it specifically did *not* involve plate tectonics. The environmental changes preserved in the strata at Ross River were simply too quick for slow-and-steady plate tectonics to have played a role. Nonetheless, they had another clue from the Ross River rocks that was difficult to explain *without* plate tectonics. In addition to sudden changes in the environment as recorded in the drastic swings in carbon isotopes at Ross River, Swanson-Hysell and Maloof had also conducted paleomagnetic sampling and found another clue in the latitude of Australia at that time—or we should say *latitudes* plural. Typically, the paleomagnetism of one stratigraphic section of sedimentary layers yields one answer, one expected "paleolatitude" for Australia at that age. However, the Ross River strata yielded *multiple* paleolatitudes, implying that Australia had *changed* its paleolatitude over a short amount of time, as if continental drift had become a continental race.

What's more is that the changes in paleolatitude occurred in lockstep with the changes in the carbon isotopes. The paleolatitudes preserved in the layers above and below (before and after) the environmental apocalypse agreed with each other, and both put Australia in the subtropics at the time those layers were deposited. But the paleolatitude deduced for the time of the temporary apocalypse put Australia squarely in the tropics. In terms of numbers, a 34° shift (almost 2,500 mi/4,000 km of motion) was implied—and not just once, but twice, thus suggesting Australia had oscillated back and forth between entirely different climatic belts. Although this was the sort of significant change of location that plate tectonics is known to be able to accomplish, as Maloof and Swanson-Hysell had suspected, there were several problems with the plate tectonic explanation.

First, the timing just didn't work. Plate tectonics has a speed limit. The fastest-known example of continental drift is India's sprint toward Eurasia. Although India covered a comparable amount of ground during its approach, its fastest burst of speed lasted about 20 million years. The two back-and-forth latitude shifts of Australia, however, had only taken a few million years each. Plate tectonics, even at its fastest, was simply too slow to explain the shifting paleolatitude of Australia.

There's a good theoretical reason why plate tectonics has a speed limit. Oceanic crust, even though it can be subducted, is still thick and must first be bent and flexed significantly downward in order to be subducted. If the oceanic crust is weaker for some reason, it is of course easier to bend. But generally the plate tectonic speed limit, around 20 centimeters per year, is set by how quickly a subducting slab can be bent.[13] At its top speed, India achieved about 16 centimeters per year, approaching the theoretical tectonic limit, which is, compared to normal plate speeds akin to fingernail growth, more like that of faster hair growth. By comparison, the motion of Australia as recorded in the Ross River strata was 80–100 centimeters per year, which is about the speed of a fast-growing lawn and thus so unreasonably fast as to cast doubt that plate tectonics could have been involved. And it wasn't just the speed of the continental motion that was incompatible with plate tectonics. It was also the peculiar style of motion.

Plate tectonics moves continents by opening and closing oceans. In other words, there's quite a bit of construction and demolition involved, almost as if you had to construct a new house and demolish your old one every time you wanted to move. The geological record should preserve some evidence of this construction and destruction.

Of course, in a book about the supercontinent *cycle* I'm not saying that such continental moves are permanent. But they certainly have some longevity, with continents staying put (relative to other continents at least) for tens, if not hundreds, of millions of years before moving again. However, the implied latitudinal shift of Australia was only ephemeral, and it was followed by an equal and opposite shift that returned Australia more or less to where it had begun. By comparison again with the example of the Himalayan tectonic convergence, India broke away from its ancient position among the southern continents of Gondwana and transited across the equator to assume its present position in the northern hemisphere in Eurasia. India's collision with Eurasia first started to cause serious uplift of the towering Himalayas about 20 million years ago. Believe it or not, but the collision is still underway, as the Indian continent is still being thrust underneath the much larger Eurasian continent. Although the India-Eurasia collision will certainly not last forever, its suture (the collision zone) is very unlikely to reverse anytime soon. Even given the relative speed of the India example, plate tectonics is simply too slow and unlikely to reverse course in a manner to explain the oscillation of Australia into the tropics and then back to the subtropics.

Swanson-Hysell and Maloof knew that if plate tectonics couldn't explain the collection of interdisciplinary observations, something else must. Another viable hypothesis to test was a much lesser-known process called true polar wander.[14] True polar wander, simply referred to as "wander" from now on, is the wholesale rotation of the solid shell of the earth (mantle and crust) about the liquid outer core that resides below the mantle. The reason wander occurs is because Earth is a rotating object that is known to move its mass around. As the earth redistributes its mass over time, wander finds a way for the planet to spin about a new, more stable axis of rotation. It is not that the axis of the planet is changing in space, it is that the planet reorients itself with respect to that stable celestial spin axis. So

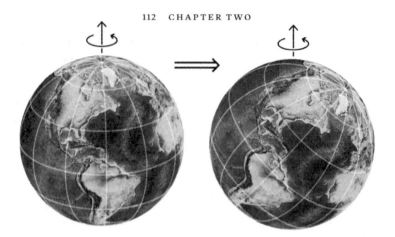

Figure 25. Cartoon of what a ~45° true polar wander rotation would look like from space were one to happen today: before (*left*) and after (*right*) wander. The arrows at the top denote the location of the spin axis. During wander, the whole solid Earth rotates around the liquid outer core to find a new geographic location for more stable rotation.

a satellite situated above the North Pole with GPS would not detect a change in the location of the Earth's spin axis: the spin axis would still be pointed directly toward the satellite. But, over time, the GPS would detect the rotation of the crust and oceans around an axis located somewhere along the equator. After wander happens, the geographic location of the spin axis is a new place relative to the continents because of the planetary reorientation (fig. 25). Again, wander occurs because Earth's solid body is dynamic: plate tectonics and mantle convection are constantly redistributing the mass of the earth. In order to respond to these changes over time, the whole solid shell of the planet (crust and mantle together) must rotate around the liquid outer core in order to stabilize Earth's rotation. But these big shifts aren't happening all the time because the rising and falling masses inside the mantle due to convection and plate tectonics can at times be more-or-less balanced around the world and thus stabilize Earth's spin axis.

To visualize wander in action, just watch a basketball spinning on someone's finger in slow motion. Believe it or not, the slight imperfections in the basketball's distribution of mass make it experience wander. As the ball starts to spin, you will see that the seams of the basketball start to shift from their random initial orientation. As the ball continues to spin and the locations of the seams shift relative to the spin axis (i.e., the finger), you will see that the "mass deficiencies" (i.e., where the seams meet up) will end up at the polar spin axis (i.e., the finger) and the "mass excesses" (i.e., where the seams diverge and there are fewer seam indentations) will end up on the "equator" (i.e., 90° away from the finger). Because the Harlem Globetrotters use a red, white, and blue basketball, you can clearly see the stabilization of the patterns over time. Definitely check this out on You-Tube (https://youtu.be/QRImf9DQwmI), because this is the easiest way to witness wander in action. The basic principle at play here is the conservation of angular momentum. As another example, think of ice skaters, who pull their arms in closer to their bodies (bringing mass closer to their spin axis) in order to speed up the efficiency of rotation. The mass of any object can be redistributed to make its rotation optimal.

Coming back to Earth, the mass excesses and deficiencies are caused by the dynamics of both rising mantle plumes and sinking subducting slabs. Because the structure of the mantle is actually quite complicated, it's wrong to think of rising plumes or sinking slabs as strictly either mass excesses or deficiencies. Their effects on Earth's rotation and thus wander can change as plumes and slabs encounter different layers of the mantle. Suffice it to say for now, it's complicated. But the basic idea is that the dynamic motions of plumes and slabs are constantly redistributing mass within the planet. The most sizable shape-shifting events of plumes and slabs in Earth history likely have caused the planet to tip via wander in search of a new stable spin axis. Thus although Australia's round-trip journey in and out of the tropics was likely directly related to wander, it is the long-term effects of plate tectonics that redistribute mass on a global scale, reaching a tipping point that triggers wander.

Plate tectonics and polar wander are thus the two ways that continents can move. However, since the plate tectonic revolution, wander has often been ignored in the study of tectonics. Before the plate tectonic revolution, wander was seen as a viable alternative to continental drift as a mechanism to migrate the continents between different climatic belts. Wander was invoked, for example, as a rival explanation for how ancient coal deposits are now found in polar Antarctica and ancient glacial deposits are now found in subtropical India. Now we know that plate tectonics and wander occur simultaneously, as proven by precise satellite measurements. Plate tectonics causes the plates to shift relative to each other, but wander is a motion shared by all the plates as the entire solid shell of the planet (including the thick mantle!) moves in unison. But when scientists proved continental drift by comparing apparent polar wander paths of continents following the breakup of Pangea (fig. 20), they discarded the rival hypothesis. They threw the baby out with the bathwater! Unfortunately, some of that attitude still lingers today, with much suspicion still surrounding the possibility of wander in the geologic past.

Assumptions that simplify solving a problem are absolutely critical to science, as nature is very complex and our minds and understanding always have their limits. Indeed, it's not an overstatement that every scientific advance is made by assuming some extra complications can be ignored, at least for the moment. But scientific progress only occurs when, standing on the shoulders of the giants that came before them, subsequent scientists question some of those assumptions. It's not that giants make mistakes (although they do that too!), it's more that scientific giants are creatures of their times. Yesterday's simplifying assumption that enabled progress is now found to impede progress. In reality, plate tectonics and polar wander are not mutually exclusive. What Maloof and Swanson-Hysell were doing was challenging the assumption that wander could be ignored as it largely had been during the plate tectonic revolution.

The abrupt changes in environment and latitude evidenced in the strata at Ross River were too fast for plate tectonics, but, on the other hand, wander might just do the trick.

• • •

Even as confusing as the Ross River data were, Swanson-Hysell and Maloof were wisely approaching the riddle as if it were a scientific crossword puzzle. They weren't just relying on one type of data or one area of expertise but on several, and the interpretation of each data set was entirely independent of the others. So, using four data sets, the student-professor duo were conducting a classic case of a scientific crossword puzzle. As confident as they might be in any one of the given data sets, the final interpretations of all the individual data sets had to be internally consistent for the wander hypothesis to be truly viable.

Wander moves continents very differently than plate tectonics does. Plate tectonics moves continents *relative* to each other and slowly. Wander, on the other hand, moves all the continents *in unison* (the whole solid shell of the earth in fact) and at speeds as fast as the fastest-moving tectonic plate, and potentially even much faster. Like plate tectonics, polar wander also has a speed limit, but it's much more lenient, more akin to the breakneck pace of the German autobahn than the US interstate highway system. But aside from this potentially significant difference in speed between plate tectonics and polar wander, the most testable aspect of the wander hypothesis—its hallmark—is surely its global reproducibility. Because wander is a wholesale rotation, this coherent motion should be shared by all continents. Studying with Hoffman during his own PhD research at Harvard several years earlier, Maloof had discovered a large magnitude (~50°) rotation of northern Laurentia.[15] If wander had occurred, the same magnitude rotation should be found in Australia. So off they were to the outback.

And indeed that's exactly what they found: an approximately 50° shift in the position of Australia relative to the magnetic North Pole.[16] (Given Australia's position on the globe at the time of the large rotation, the continent experienced of latitude change of ~34°, as we mentioned, with the extra angular motion being expressed in the rotation of Australia.) At face value, it would seem like resounding support for the wander hypothesis. But as science would often have it, the Australia results were not exactly clear cut. Even though the paleomagnetic data from the carbonate rocks indicated a large shift between two paleomagnetic directions, one of the directions could not be unambiguously proven to be reliable. As the carbonate rocks had been tectonically tilted, there was the possibility that heat and fluid from that subsequent tectonic event had remagnetized the rocks. Nonetheless, the data weren't proven to be unreliable either. In their scientific paper at the time, Swanson-Hysell, Maloof, and colleagues described the wander interpretation as taking a "rose-colored glasses view": if one wanted to interpret the data as supporting the wander hypothesis, one could; but it wasn't a definitive test—all the more reason to take a scientific crossword approach and not rely on one line of hypothesis testing alone.

So what about the carbon isotopes? Since the carbon cycle is global in nature, one must consider the distribution of all continents at that time, not just Australia. One must also focus on regions most important to the carbon cycle. On Earth today, the mighty Amazon River must come to mind. Running from the Andes Mountains in western South America all the way to the ocean on the east coast, the Amazon River not only crosses an entire continent, it does so in the tropical weathering belt near the equator. In doing so, the Amazon River delta buries a lot of organic carbon in the massive amount of sediment it deposits. With an Amazon River, the fraction of organic carbon being buried is high. If wander were suddenly to occur, however, and shift the continents such that the Amazon River delta was no longer on the equator but near the pole, then the fraction of organic carbon being buried would be greatly reduced. Move the Amazon off the equator, and you change the global carbon cycle.

This is precisely what might have happened about 800 million years ago, when the Ross River section was deposited. Although the Andes Mountains and the Amazon River did not exist yet, other towering mountains flanked other mighty rivers—namely the familiar Grenville Mountains, running down the spine of supercontinent Rodinia, and the many rivers that drained from them! Similar to the modern Andes, the Grenville Mountains straddled the equator. In the tropics, organic material is abundant and the sediment to bury and preserve that organic carbon is also not in short supply due to the abundant precipitation that constantly weathers rocks. And, also like the Andes that run down the entire coastline of South America, the Grenville Mountains spanned the entire coastline of North America. Grains of sand weathered from the Grenville Mountains in the eastern United States have been found in the far-flung northwest corners of Canada. In other words, Laurentia indeed had thunderous rivers, as long as the Amazon, running off the Grenville mountain chain and across the entire vast continent.[17] But unlike the Andes, a mountain chain formed by subduction and therefore located along the edge of a continent, the Grenville Mountains were a bit more like the Himalayas: mountains made by the collision of continents. Such mountains lie in the *interior* of large supercontinents. Therefore, whereas snowmelt and rain from the Andes only source one Amazon River running across one continent, the Grenville Mountains would have had several large rivers running across at least two continents.

So with Laurentia and its Grenville mountain chain straddling the equator, recall from Hoffman and Li's maps of Rodinia that Amazonia itself and other continents surrounding Laurentia would have also been in the tropics. This meant it was a good time for life in the oceans, because the combination of tropical weathering and Amazon-like rivers transported nutrients like phosphorus, once trapped in rocks, into the oceans in great abundance. Photosynthesis requires nutrients as well as sunlight. This extra fertilizer would have stimulated life, and therefore this was the situation both before and after the deposition of the Ross River strata. But the carbon isotopic value of the oceans suddenly and temporarily plunged to lifeless values akin to volcanic emissions during deposition of these strata. So what changed? And what changed it back? Could wander explain this?

Combining paleomagnetic data from Laurentia from Maloof's PhD research with that from Australia from Swanson-Hysell's predicted what this approximately 50° rotation would have done to the arrangement of the continents at that time. The resulting effect of the rotation of the continents predicted a profound change in the way the carbon cycle had been operating to that point. Before the rotation, the Grenville mountain chain straddled the equator, but with the rotation, this favorable orientation for photosynthesizers was no longer. Large sectors of the Grenville Mountains and the Amazon-like rivers that flanked them would have shifted from the humid tropics into the arid subtropics, thus shutting down many of these organic carbon factories. And it wasn't just Laurentia but most of the continents surrounding it in the Rodinia supercontinent that changed from a predominantly tropical configuration to one with many continents in the subtropics or even higher latitudes. Indeed, this same rotation has since been confirmed in South China.[18] The world was suddenly without tropical weathering efficiently delivering nutrients into the oceans to fuel photosynthetic bacteria and algae; it was also without the weathered sediment load with which to bury the little organic carbon that was being produced. Both of these effects—less organic carbon generated and less of it buried—conspired to prevent the concentration of the heavier and photosynthetically unfavorable carbon-13 in the oceans. The global carbon cycle thus plunged to carbon-12-rich values akin to volcanic carbon emissions left to their own devices.

But wander could also explain why this relatively lifeless world only lasted so long and why it reversed just as fast. Whereas plate tectonic motions like India colliding into Eurasia are not likely to reverse until millions and millions of years later, wander, on the other hand, can. As mentioned before, subducting slabs and mantle plumes, as they sink and rise respectively through the mantle, can actually change from being mass excesses or mass deficiencies. This is because the mantle is not homogenous but actually layered into a very stiff lower mantle and a less stiff upper mantle. When a dense subducting slab starts to sink through the upper mantle, it creates a mass excess that for stability wants to be shifted toward the equator; however, once it sinks into the super-stiff lower mantle, the same slab then becomes a mass deficiency that wants to be shifted toward the pole.[19] The same reversal of effect on wander is also true for buoyant rising plumes, but in the opposite sense.

Another reason for wander being able to oscillate back in the direction from whence it came is based on the nature of the lithosphere, the outermost solid layer of the planet that is involved in plate tectonics. The lithosphere contains the entire compositional layer of the crust, and even a bit of the mantle too. Because of its distinct physical properties, the lithosphere may deform elastically, that is, it may behave like a rubber band which recovers its original shape once stresses are removed. Because Earth is rotating, it isn't a perfect sphere, but more like an oval that bulges at the equator. As a result, when wander rotates the whole crust and mantle together, the lithosphere has to stretch to accommodate changes in latitude. Because of its elasticity, the lithosphere may have a "memory" for its previous shape, thus later snapping back into its prior shape, before wander occurred.[20] In short, as easily as wander can explain the motion of the Grenville Mountains and the surrounding continents of Rodinia temporarily out of the tropics, it can also explain the return trip. And once wander oscillated back to its original pole position, the same continental margins flourished with photosynthesis and organic carbon burial again, thus restoring the carbon isotope values to their previous levels. Wander can be a round-trip ticket.

Why have we spent so much time discussing polar wander in a book about the supercontinent cycle and, to a large degree, about plate tectonics—particularly, when I have also mentioned that many scientists continue to think of wander as a hypothesis discarded when continental drift was proven? The short answers are that current plate tectonic theory is insufficient to understand the supercontinent cycle and that wander is a critically missing piece of that puzzle as it offers clues about mantle convection that plate tectonics currently doesn't. Because of its importance in our larger story about predicting the next supercontinent, the feedback between plate tectonics and mantle convection is paramount. Hence, we must explore all the clues we have about mantle convection inasmuch as we explore the plate tectonic motions of the continents. The hope is that after seeing the multiple lines of evidence from the Ross River strata that can be explained by polar wander, but not plate tectonics, you agree that wander is a force to be reckoned with, one that we must better understand. In addition to this intellectual appeal, I must also appeal to your emotions for you to fully understand my complete commitment to better understanding wander and the clues it can offer us about the supercontinent cycle. I must finally finish telling the story of how I lost half my right thumb.

•   •   •

For the purposes of the thumb story, it is most important to know that dolomite is typically harder and sharper than limestone. The already worn soles of our once new boots could attest to this, as can the appearance of the jagged weather-resistant mountains of the Italian Alps—the Dolomites.

Swanson-Hysell and I had been camped for almost a week already at the Ross River section. We'd already been up and over the section in great detail having sampled quite regularly for carbon isotopes and measured the thickness of the sedimentary rock layer compositions. It was finally time for the paleomagnetic sampling . . . drill baby, drill! Sampling rocks for paleomagnetism, we use a chain saw that has been modified into a portable rock drill. Instead of a chain for cutting wood, the rotor spins a diamond-tipped drill bit that cuts rock through friction. As diamond is the hardest mineral known, it wins any battle of friction. (No, these are not the rare, naturally occurring gemstones, but industrially fabricated diamonds.) Since friction creates a lot of heat, we also have to pump water to the tip of the drill bit to cool it down so it doesn't thermally expand and crack. In the Australian outback, water is scarce. There are no rivers or streams—they pop up only temporarily during flash floods after rainstorms, but then they dry up (we actually searched these dusty creek beds for the best firewood). This is a land where farmers pump groundwater from boreholes. So we had to fill up our water jugs every time we filled up our gas tank. Water, sometimes before petrol, was what would bring us back to civilization. But on that fateful day, we would be returning to the city for a whole new reason.

When you're drilling, you are carrying a bit more gear than normal. But we didn't think twice about it. It's what it took to get the job done. We were working on a hill slope that was steep enough, with unsteady crags here and there, that it's probably not a place you'd take a class of students. But we didn't think twice about working on that slope with just the two of us for fieldwork. It's just what it took to get the job done. I came to a tricky spot where I needed to use all fours to climb up on to the next ledge. I hoisted the drill and gas can I was carrying up on to the ledge I was about to climb. I got my hands in position and was putting my feet in position when I noticed the rock in front of me was shifting.

The whole rockface was coming loose. I fortunately had space off to my right and I was able to pivot out of the way as it broke free. The cracking slab was torso sized and torso height. I pivoted on my right foot, sweeping my left foot and hand out of the way of the falling slab. I thought I had made it clear. But now the boulder was tumbling past me and picking up speed.

Luckily, Swanson-Hysell had wisely taken a different pathway up the slope and was slightly offset from me. By the time the rock whirled past him, it had picked up enough speed that it probably would have killed him, or at least knocked him unconscious. But instead, it just careened on by. In the clear, but concerned for me, Swanson-Hysell hollered up to me, "Are you OK??!"

I thought I had gotten out of the way just in time. And I nearly had. But as I considered how to answer Nick's question, I realized that my right hand hadn't got out of the way quite fast enough. As I had pivoted away from the falling rock, my right thumb had gotten pinched underneath it. With the first surge of adrenaline hitting my veins, I answered Nick loud and clear and very matter-of-factly for the circumstances.

"I just lost my right thumb," I announced.

"OK," Nick said, "Stay where you are, I'll be right there."

Nick clamored up the tricky dolomite crags to get to where I was. Once he got there, we assessed the situation. Contrary to my initially bold statement about having lost my right thumb, we realized the situation wasn't quite as bad as that. I'd only lost part, maybe half, of my right thumb. Nick asked me if I knew where the rest of it was. I tried not to laugh because I knew he was serious and was right to ask. But whatever had been lost, had been lost. The boulder might have even taken it with it. We looked around for a minute before giving up the search and turned our focus to the problems ahead. We didn't have much in terms of clean cloth wraps, so without hesitation, Nick took his shirt off and ripped a bandage-sized strip and tied up my hand so it would be safe for our descent to the vehicle.

As we started the descent, Nick was in front of me and kept turning back to spot me. I could tell he was shaken, and I would have been too if it had been my friend who was hurt. But with the certainty of adrenaline, I let Nick know, "You don't need to spot me. I didn't lose a leg, just a finger. Just lead the way and I'll be right behind you." We made our way down to the campsite. Once we reached flat ground, Nick ran up ahead to prepare the truck for our getaway. We had set up our cooking table behind the hatch of the truck. With one fell swoop, Nick tossed the table out of the way and ran to start the truck. We didn't have any ice, so we used the air conditioning in the truck to keep my severed thumb as cold as possible until we got to the hospital.

Once we were in the car, it was an unsteady twenty minutes or so of off-roading before we finally hit paved road. By then, my adrenaline had started to wear off and the pain and reality of the situation began to set in. And then Nick turned our field truck into an ambulance. He passed every car on that straight road running west to Alice Springs (fig. 26). Even at this fast clip, it took us several hours to get to there.

We finally arrived at the Alice Springs ER. Still shirtless, Nick took a few seconds to put one on before rushing me into the ER. I remember feeling discouraged by how many other people were already waiting. But given the time-sensitive nature of my severed thumb, it luckily wasn't long before I was lying flat and receiving a much-needed morphine drip. Maybe it was the morphine speaking, but I recall one of the nurses asking which of my toes I would prefer to transplant to my thumb. Perhaps this question was intended to reassure me, but that wasn't exactly what I wanted to hear at the time. Fortunately, I had Nick in my corner every step of the way.

After a phone call from Nick, my parents offered immediately to fly halfway around the world. I was of course consoled by the gesture, but kindly declined as the practical benefit wasn't obvious to me and time was of the essence. Feeling the parental impulse to do something, they got on the phone with hand doctors in the United States to help get us second and third and fourth opinions on what my options might be.

Figure 26. The drive from Ross River (fig. 22; in the right side of this photo) to the Alice Springs hospital. There is a two-lane highway east of Alice Springs. The initial portion of the drive involved off-roading on a dried-up creek bed (dashed line). Image is about 44 miles (~70 km) wide. Courtesy of the public domain from https://earth explorer.usgs.gov.

The good news was that the falling boulder had spared the thumb joint. Every hand doctor that my parents spoke to at home insisted how fortunate this was, as this joint is what gives us our opposable thumbs.

The doctor in Alice Springs heartily agreed about the importance of preserving the joint if possible and certainly seemed to know what he was talking about. But the options he gave me meant a difficult decision. He would try three solutions, in successive order as needed. First, he would try to use the existing soft tissue to close the wound. If that failed (if there wasn't enough soft tissue), he would then close the wound with a skin graft from my thigh. If the skin graft didn't take, then he would be forced to amputate at the joint in order to have enough soft tissue to close the wound. And since these iterative options would all be tried one after the other while I was under anesthesia, he insisted on me agreeing to all of these options before trying any of them—including option C . . . amputation.

All the hand doctors in the United States insisted that amputation was completely unnecessary, and that amputation would limit options later if I ever wished to have a prosthetic extension. Thumb joints are very complex; with the joint intact, a prosthetic extension would be straightforward; without the joint, however, prosthetic options would be limited or less than ideal. But in order to let any of them have it their way, it would involve delaying the surgery (increasing the chance of infection) and flying halfway around the world with an open wound.

I decided to trust the doctor that amputation was only a last resort. Long story short, there wasn't enough soft tissue to close the wound (option A failed), but the skin graft took (option B succeeded). My thumb may be a bit shorter, but it's still opposable and the damage essentially unnoticeable. This happened to me in my junior year of college, even before the start of my career. But it never made me waver from my commitment to becoming a geologist. Once they heard the news, the potential professors I was courting as graduate school advisors were impressed by my resolve. Sometimes you have to risk losing a finger to follow your heart.

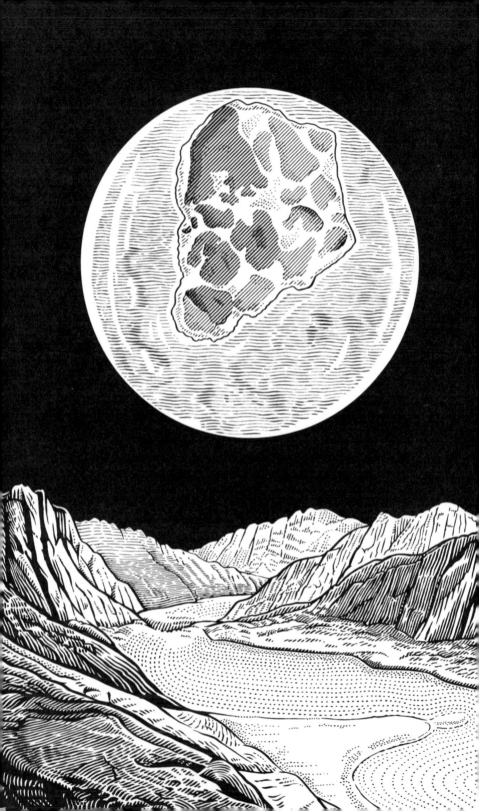

# 3

# COLUMBIA

The past is never dead. It's not even past.

WILLIAM FAULKNER

As far as convincing ourselves that the supercontinent *cycle* indeed exists, we need three examples. If just Pangea existed, it could have been chance. Two supercontinents (Pangea and Rodinia) are still in the realm of coincidence. Three supercontinents is science. In order for there to be a supercontinent cycle, we therefore have to convince ourselves of the existence of not only Pangea's predecessor, Rodinia, but also of the reality of Rodinia's predecessor.

If there was one moment that decided where I went for my PhD, it was this. When I was deciding where to go for graduate school—2006, incidentally also the year I lost part of my right thumb—scientists viewed the notion of yet another more ancient supercontinent as highly speculative. But even the remote possibility compelled me. I was already aware that David Evans at Yale had plans to test the idea of Rodinia's predecessor and he had also given presentations at conferences indicating some early clues of its existence. But after the amazing opportunity that Adam Maloof at Princeton had given me to do exciting fieldwork in the Australian outback with my friend and colleague Nick Swanson-Hysell, he was initially my first pick of PhD advisors. Then I asked him a question: Did he believe in a pre-Rodinia supercontinent?

"It's more fiction than fact," Maloof dryly responded.

I was disappointed by his hopeless outlook. Not because he had revealed to me any convincing evidence that the ancient supercontinent hadn't existed, but because he revealed that he and I approached science very differently. To do a PhD under someone's supervision, you must then agree where to place your collective bet. Based on the evidence that I had already seen, I was willing to stake my dissertation on testing the possibility of yet another supercontinent. Of course, Maloof was right to point out that the evidence was scant, and I could understand his valid, albeit disheartening, response. But the indications I had seen from David Evans's conference reports and grant applications seemed to point in the right direction even if they were still quite far from what Maloof would accept as fact. My first choice for graduate school became Yale, and I didn't even apply to Princeton.

The main early clue that had me willing to bet on the possibility of another supercontinent was that David Evans had identified its "keystone," the continent that holds the puzzle together. Pangea had Africa and Rodinia had Laurentia. Reconstruction of these supercontinents had both started with the identification of the central piece of the puzzle around which the other continents were configured. South African geologist Alexander Du Toit, Alfred Wegener's contemporary, also pondering the impermanence of the continents, was wise to look at his home continent for the most convincing clues about Pangea's existence. In Du Toit's *Our Wandering Continents*, he claims that Africa, with its distinctive shape of shoreline, fossil-rich and diagnostic strata in the Karoo of South Africa, and well-exposed Precambrian rocks, "forms the key" to reconstruct Gondwana, the large southern portion of Pangea.[1] Wegener, and even paleogeographers to this day, employ this keystone approach to reconstructing ancient supercontinents. In the last chapter, we saw how solving the Rodinia puzzle started with Hoffman arranging his cardboard cutouts around a centrally positioned Laurentia. When it came to the possibility of yet an even older supercontinent, Evans and Russian scientists had already identified a candidate keystone for Rodinia's predecessor, and it was Siberia.[2]

Based on the ages of ancient mountain belts, the pre-Rodinia supercontinent—if it indeed had existed—likely assembled about 1.7 billion years ago.[3] The details of those all-important mountain belts will be explored in due time. First, let us look at how a supercontinent breaks apart, because it also leaves telltale signs in the rock record similar to the way mountain belts leave evidence of supercontinent assembly. Think of East Africa: the only example on Earth of a continent actively breaking apart today. The geology of the East African rift valley consists of basalts and "horst-and-graben" rift basins. These are the igneous and sedimentary telltale signs of continental rifting. Basalt is a "primitive" igneous rock, meaning that its melt is directly sourced from the mantle; the easiest way to achieve this is if the mantle is close to the surface, which can occur due to stretching and thinning of the crust above. Horsts and grabens are opposite sides of the same coin. When the crust is stretched, it thins because faults allow blocks of the crust to slide away from each other. Horsts are the crustal blocks that are raised and grabens are the valleys in between. Starting around 1.5 billion years ago, and especially by 1.3 billion years ago, many continents showed the evidence of both basalts and horst-and-graben rift basins that provides the clues of supercontinent breakup at that time. But the record in Siberia was most striking: Siberia was essentially surrounded on all sides by such clues of continental dispersal. If Rodinia's predecessor had existed, it had most likely been a supercontinent with Siberia at its core. With the keystone in hand, those like Evans and myself were willing to place our bets on testing this hypothesis. But it would take a lot of work to go from the perception of fiction to fact.

● ● ●

There is a reason I have referred to this next oldest potential supercontinent as "Rodinia's predecessor." There is confusion among scientists what to call it: Nuna or Columbia. From a scientific point of view, I find the Columbia vs. Nuna battle insufferable as it has zero impact on the science, but I'll share some of the details nevertheless.

In 1997, Canadian Paul Hoffman used the term "Nuna" (an Inuit word meaning "all land," for the lands bordering the northern oceans and seas of North America) to refer to the larger continent of Laurentia that was suspected to be contiguous with Baltica.[4] Thus, the Nuna camp says that Hoffman was first on the scene to plant a flag for the name of the supercontinent that assembled some 1.7 billion years ago. But those in the Columbia camp would point out that Hoffman was only specifically referring to a connection between a small handful of continents (Laurentia and Baltica) and did not offer a global-scale supercontinent reconstruction. In 2002, the term "Columbia" was coined referring to the Columbia River region of the United States' Pacific Northwest in the context of it being part of an ancient supercontinent.[5] Then, in the same year, a research team led by Professor Guochun Zhao of the University of Hong Kong published the first paper to present a supercontinent reconstruction for this age.[6] Zhao and his team chose to perpetuate the name Columbia. Why the Columbia namesake comes from a region with little to no rocks 1.7 billion years old is strange. And Nuna arguably has precedence on its side simply given the dates of publications. But the two 2002 papers have now *each* been cited over 1,300 times, and Columbia clearly isn't going away any time soon. Such popular success arguably demands respect.

And so, back and forth in the scientific literature, each camp continues to publish its papers promoting the name that it prefers, and those that are indifferent and caught in between will verbosely refer to the supercontinent cycle as "Columbia/Nuna" or "Nuna (a.k.a. Columbia)." Interestingly, the namesake stalemate is arguably a manifestation of the current political confrontation between the United States and China. Western scientists more often use Nuna and those in Asia (most publishing currently comes from massive

China) tend to prefer Columbia. How will this stalemate be resolved? Will it be? The arguments of precedence or priority are obviously not as clear cut as reverting Ayers Rock to Uluru, which even still is officially registered by the Northern Territory Government of Australia as Uluru/Ayers Rock. From an outsider's perspective, you're probably thinking, "just pick a name already and get back to the science." Conveniently, my colleagues and I might have recently found a solution—and a scientific one at that. But we are getting ahead of ourselves.

•　•　•

In our Rodinia chapter, we only touched the tip of the iceberg of Paul Hoffman's contributions to ancient supercontinents while he was at the Geological Survey of Canada. Ask a random geologist about Hoffman's scientific legacy, and typically the words "Snowball Earth" will be what you hear. But it's easy to forget that the prolific Hoffman had an entire world-class career before moving to Harvard and unearthing the Snowball Earth hypothesis. As a supercontinent devotee, I'd say Hoffman's contributions are second perhaps only to those of Alfred Wegener himself. While Hoffman's prophetic vision of supercontinent Rodinia has spurred on the decades of research since, this was only his final tour de force published at the Geological Survey of Canada. Long before this, Hoffman had already established supercontinents as the forefront of research into ancient plate tectonics. First, Hoffman became known for the "United Plates of America."

Every continent is bestowed with at least some ancient rocks. But it's the overwhelming number of old rocks that makes North America special. And, as luck would have it for Hoffman, Canada has most of the old rocks compared to the United States. Thus, at the Geological Survey of Canada, Hoffman was working at the perfect place to make a big difference in developing our understanding of ancient plate tectonics. Also as luck would have it, during the most recent ice age—the "last glacial maximum" 31,000 to 16,000 years ago—most of Canada was bulldozed by grinding glaciers. These icy forces of nature are extremely efficient at erosion compared to weathering from rainwater. The ancient rocks of Canada, some of which had been covered by younger rocks forming on top of them, have thus since been beautifully exposed for geologists to walk over and access easily without drilling. In a word, Canada is a treasure trove for studying ancient tectonics and for the search for yet older supercontinents.

Hoffman would eventually make his mark by synthesizing the Precambrian geology of all of North America. But even the stories of such giants tend to start with humbler beginnings. There's no need to give Hoffman's life story before he became a geologist. That gripping tale, which includes first-time marathoner Hoffman placing ninth in the 1964 Boston marathon, is told well by Gabrielle Walker in her page-turner about Snowball Earth.[7] Our story starts in the humblest beginnings of any scientist: the PhD thesis. With an interest in sedimentary rocks and old rocks—after all, Hoffman hails from Canada—he chose to pursue graduate school under the supervision of Francis John Pettijohn at Johns Hopkins University. Pettijohn had just written timely textbooks on sedimentology in the geologic record. But Hoffman's reasons for his decision didn't stop there. Franklin and Marshall College, near Johns Hopkins and where Hoffman had been teaching before enlisting in graduate school, was running a program with the Geological Survey of Canada to send students into its field camps. So young Hoffman would not only be able to learn sedimentology from one of the best experts, he also got to travel far up north to his homeland to visit some of the best old rocks. His interests would take him even farther north than his upbringing in Ottawa. Hoffman was headed to the tundra.

•   •   •

It's amazing we've made it this far without discussing the geologic timescale, the most fundamental framework of geology (fig. 27). It's how geologists tell time. And when that history spans thousands, hundreds of thousands, millions, even billions of years, having a system for telling time is critical. Sure, we could only deal in numbers; but not only is that boring, it's also impractical. It would be like counting the 365 days of the year without referencing the four seasons, the months, or the weeks. Instead, then, Earth history is broken up into a hierarchy of longer time intervals subdivided into shorter and shorter increments—but only as the natural course of events on Earth allow. Spans of time where little happened on Earth are characterized by fewer, less frequent time intervals; spans of time where Earth was a flurry of environmental and biologic changes are characterized by more frequent time intervals. In a word, the geologic timescale is a yardstick of global change.

In our next chapter, we will go back into deep time and discuss the largest, most fundamental chunks of time in the hierarchy of the geologic timescale: eons. But because eons are so long (on average, 1 billion years long!), they are typically subdivided into three eras with prefixes like Neo- (new), Meso- (middle), and Paleo- (ancient). For example, the Proterozoic eon, with at least two supercontinents, is subdivided into the Neoproterozoic (the time of Rodinia), the Mesoproterozoic (the time of Nuna/Columbia), and the Paleoproterozoic (the lead-up to Nuna/Columbia, the time Paul Hoffman was concerned with as he headed far north for the first time in his life). But such eras are further subdivided into one of the more expressive intervals of time of Earth history: periods. In the Mesozoic era, for example, there are the three well-known Triassic, Jurassic, and Cretaceous periods. Although the Jurassic was made the star period of the era by Michael Crichton and Steven Spielberg, dinosaurs became much more diverse in the Cretaceous period. Anyway, the main point is that each period of Earth history has a unique character and denotes a particular time in Earth's evolutionary clock.

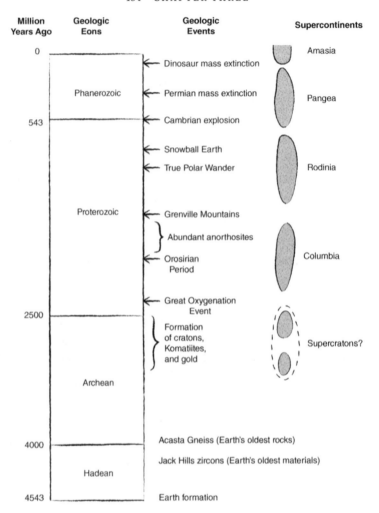

| Million Years Ago | Geologic Eons | Geologic Events | Supercontinents |
|---|---|---|---|

Figure 27. Geologic timescale with events discussed in the text.

Our story, as young Hoffman headed into the tundra, concerns a different period in a much older time: the Orosirian period of the Paleoproterozoic era about 2 billion years ago. *Orosira* (Greek) means "mountain range," and this is exactly what Hoffman was far up north to study: ancient mountains that could tell him about ancient plate tectonics—and maybe even something about ancient supercontinents. The Orosirian period was thus defined by mountain building, or orogenesis, that occurred on virtually all continents at that time. More than twenty orogens have now been discovered during the Orosirian.[8] And since making mountains is a natural consequence of continents colliding, the Orosirian is thus a natural time to search for the possibility of an older supercontinent, the putative predecessor of Rodinia.

•   •   •

Hoffman's approach to studying ancient mountains may surprise you. Remember that he had chosen to study for his PhD at Johns Hopkins under Pettijohn, an expert in sedimentology. This subfield of geology studies what sediments can tell us about the rise and fall of sea level as well as the type of environment and climate in which sedimentary rocks were deposited. So what does sedimentology have to do with mountains? Hoffman would in fact spend relatively little time directly studying ancient mountains. He spent most his time studying the sediments that got shed off the mountains into the basins around them. Why?

Sedimentary rocks are the closest thing to the archives of Earth history (no disrespect to invaluable igneous and metamorphic rocks). But because of their layer-by-layer nature, sedimentary rocks keep a remarkable record of events as they happen. Sure, we can date igneous and metamorphic rocks, but billions of years ago even the most precisely determined rock ages have uncertainties as large as millions of years. On the other hand, each layer of sediment deposited on top of the last one is unassailably younger—and *relative* time, known as the principle of "superposition," is one of the most fundamental concepts in geology and a huge advantage to telling time in sedimentary rocks. But Hoffman and his colleagues also had tricks to getting *absolute* time and precisely dating the sedimentary rocks in exactly the same manner as one can date igneous and metamorphic rocks—with uranium-to-lead (U-Pb) geochronology as before.

Not only do mountains shed sediments, they are also often known for their volcanoes. Before two continents can collide, the potentially large ocean separating them must be closed. Think of Australia on its current collision course for Eurasia. Someday those continents will collide, but until then, a lot of oceanic crust between them in Southeast Asia must be subducted. Being underwater, oceanic crust gets more and more hydrated over time as it incorporates more seawater into the pore spaces of its basaltic crust. Thus, when oceanic crust is subducted beneath a continent, the descending slab starts to dehydrate as it reaches hotter temperatures deeper in the mantle. The flux of water out of the slab that eventually makes its way into the overlying continent is a perfect and simple "just add water" recipe for creating the magmas that form granites at depth and also feed volcanoes at the surface. The ring of fire that encircles the subducting Pacific Ocean is where most modern volcanoes on Earth are found. These are "magmatic arcs," and there's certainly one in Papua New Guinea as Australia races toward Eurasia and all intervening oceanic crust in its path is subducted.

Importantly for Hoffman's team, volcanoes are prone to repeatedly erupt not just molten lava but also a lot of volcanic ash. And it travels very far. The "supereruptions" of the Yellowstone Caldera, for example, scattered ash across much of the continental United States, as far as Louisiana—a respectable domestic flight![9] As the Orosirian mountains were forming, the volcanoes of their magmatic arcs were shooting plumes of volcanic ash into the sky that settled in the sedimentary basins at their feet. The same mineral forming in magma chambers that make igneous rocks possible to date precisely with U-Pb geochronology, zircon, is also aerodynamic and takes flight with volcanic ash. As Hoffman and his students investigated the sedimentary basins layer by layer, they would occasionally find a layer of volcanic ash marking its moment in time between two layers of sediment. They would collect big bags of ash, bring them back to the lab, separate out all the zircon grains they could find, and then date the zircons. Not only could Hoffman tell relative time with sedimentary rocks, he knew quite precisely their absolute ages too.

•  •  •

U-Pb dating of ash layers allowed Hoffman to tell time, but how could he tell that the mountains were rising and leading up to continents colliding? This evidence would have to come from the sedimentary rocks themselves. One of the techniques that Hoffman had picked up from his esteemed supervisor Pettijohn was the usefulness of "paleocurrents," or evidence of ancient water currents preserved in the sedimentary rocks they deposited. Think of ripples in the sand as the waves recede from the shore. The crests of the ripples are oriented perpendicular to the wave direction. There are also ripples preserved in river sediments, but since the flow of water is different than it is in ocean waves, so too do their ripples differ. Ocean waves oscillate back and forth, making the two sides of the ripple crest symmetric. In a river, however, where flow is constantly in one direction, ripples are asymmetric, that is, they "lean" toward the direction the river is flowing. Most importantly, such paleocurrents allow the geologist to detect which direction water currents carrying sediments had come from.

True to his marathoner form, the young Hoffman took his task of measuring paleocurrents in the sedimentary rocks he was studying quite seriously and went the extra mile, if you will. By the time the field work for his thesis was complete, Hoffman had taken a total of 8,000 paleocurrent measurements, mapping their orientations using a magnetic compass. (Most exceptional studies deal in hundreds of data, not thousands.) By his own admission, Hoffman mentioned that his number of measurements made might have been "excessive" and that "1,500 measurements from a given formation yield no more useful information than 500."[10] But Hoffman had no regrets. Ultimately, they served to "make the paleocurrent maps appear more convincing," and such a level of certainty in the world of science, if attainable, is a holy grail worth questing for.

From Hoffman's detailed paleocurrent maps it became clear that a fundamental change had taken place from one sequence of sedimentary rock to the next. Initially, in the older sedimentary rocks, the paleocurrents had indicated what one might expect: the sediments had been sourced from the nearby continent on which the basin had formed. But then, in the geological blink of an eye, everything changed. Suddenly, in the younger sedimentary rocks, the paleocurrents indicated just the opposite: they had turned around and thus could not have deposited sediments shed from the nearby continent—they had to come from somewhere else.

Although we think of North America, or Laurentia, as one continent today, this has only been the case since the Orosirian period about 2 billion years ago. That is, before the United Plates of America of Orosirian time, there were many independently drifting plates. These ancient continental nuclei that made up Earth's relatively young and small continents are referred to as "cratons." Laurentia is composed of about five or six cratons that all collided in Orosirian time. Hoffman had made his paleocurrent measurements on the Slave craton. But the sudden change in the provenance of the sediments told him that another craton was quickly approaching. Eventually, with the help of graduate students, Hoffman would date this collision between cratons to 1.97 billion years ago.[11] Our understanding of the United Plates of America was beginning and the Slave craton, like Hoffman himself, was there for the very start of the action.

•   •   •

Researchers have made a lot of progress since Hoffman's United Plates of America and have expanded our concept of Rodinia's predecessor well beyond Laurentia—so far, in fact, that we must travel once again to the Australian outback. Another giant in the field of supercontinents, Zheng-Xiang Li, like Hoffman, has contributed to our understanding not only of Rodinia but also of its predecessor. In the last chapter when discussing Li's efforts to understand Rodinia, I neglected to mention that I was fortunate enough to have worked with Li for three years as a research fellow under his inspiring supervision. But it wasn't just me. As with his research on Rodinia, Li investigated Rodinia's predecessor. Li formed an interdisciplinary research team comprised mainly of early-career researchers, about fifteen to twenty of us over the five-year grant. I viewed it as the opportunity of a lifetime, a chance to initiate collaborations with colleagues that will continue to bear fruit for years to come.

A young graduate student in Li's group would make one of the most significant discoveries of this project in pursuit of his PhD. Following Hoffman's footsteps, and under the expert supervision of Li, Adam Nordsvan used the sedimentary rocks of Australia to test where the outback might have been in this ancient supercontinent. Although Li's group hailed from all over the world—the United States, China, Germany, France, Russia, Egypt, and Italy—Nordsvan was one of the few Aussies representing our host country.

Nordsvan and Li had the idea that they could "fingerprint" the sedimentary rocks of Australia as a way to identify where they had come from. Were the sediments in fact locally derived from nearby sources in Australia, or was their origin story more exotic? Methods of "fingerprinting" had come a long way since the days of Hoffman's PhD. For his PhD, Nordsvan would rely predominantly on a now popular technique called "detrital zircon provenance." This basically involves imagining the life of a zircon from when it first crystallizes to everything that may happen to it thereafter until it is deposited in a sediment. So let's give an example. A zircon crystallizes deep in a magma chamber. Quite commonly, as the magma cools and crystallizes, the zircon becomes one of the minerals in granitic rock. But someday later, the granites can reach the surface when the overlying crust is stripped away by uplift and erosion over millions of years. Arriving at the surface, the granite can be weathered and turned into sediment. The zircon is set free, becoming a "detrital" zircon as it is transported by a river toward the ocean. Along with other sediments, the zircon finds its way into a sedimentary basin, gets deposited as a sediment, and is eventually preserved as a sedimentary rock. The task for Nordsvan was thus retracing the transportation of the zircon back to its source, the actual exposed rock from which the zircon was plucked and became a grain of sand. In short, Nordsvan planned to date detrital zircons as a way of finding where the sediment had come from.

This method works because not all continents have the same-aged rocks. Sure, there's some overlap, because rocks are made all the time all over the world. But, as a general rule, certain age "peaks" in zircon can be attributed to only one or two continents. As an example, few places today are generating as much granite (and zircon minerals in granite) as the Andes of South America. Now, identify one or two or, even better, multiple age peaks in your detrital zircons and the chances of success increase exponentially. It becomes less likely that more than one place in the world has the same exact age peaks. Nordsvan's field area in Queensland, Australia, is referred to as the Georgetown Inlier as it appears it may have been its own tectonic entity at some point prior to its potential incorporation in Australia. That is, Georgetown is now found in North Australia, but that might not have always been the case. Comparing the zircon age peaks of Georgetown to those of its surrounding crust in Australia and elsewhere in the world could test this idea that Georgetown might have come from somewhere else.

After the hard work of dating his 605 detrital zircons, Nordsvan was elated to indeed see the presence of well-defined, dominant age peaks for his field area in Georgetown. Then came the moment of truth: to compare his age peaks with those of the local sources of North Australia and with those of more exotic locations around the world. It was immediately clear that Nordsvan would have to look beyond this small corner of North Australia to find a "fingerprint" match. Keep also in mind that, because of plate tectonics and the old age of these rocks, the nearby regions presently surrounding Queensland either didn't exist at the time (the rocks hadn't yet formed) or they were tectonically drifting independently and hadn't yet collided with Queensland. Thus, not only did Nordsvan need to look beyond his immediate surroundings in North Australia, he also might need to look well beyond Australia.

Although the match was compelling, the hypothesis it inspired— that Georgetown had in fact broken away from North America and then collided with North Australia—had to be tested further with an independent method. Thus, in addition to Nordsvan's dating hundreds of detrital zircons and using modern mathematical algorithms, he also used a trick straight out of Hoffman's playbook: paleocurrents. Just as Hoffman had done in the Slave craton, Nordsvan measured the paleocurrents of each of the sedimentary rock formations through time. But going one better than Hoffman, Nordsvan analyzed paleocurrents from the same formations he was analyzing for detrital zircons. Interestingly, Nordsvan found a 180° pivot in his paleocurrent measurements through time. Furthermore, the aboutface corresponded precisely to the change in detrital zircon provenance. And, to top it off, the two contrasting paleocurrent directions made sense with the two contrasting sources. When zircons were thought to be coming from North America, Georgetown sediments were sourced from the east; then, when the zircons shifted their provenance to North Australia, Georgetown sediments became sourced from the west. With their crossword-puzzle-like corroborating evidence, Nordsvan and Li had proven that part of the Australian outback was in fact born in the USA.

Nordsvan's discovery that part of Australia originally came from Laurentia means that these two continents were neighbors in Rodinia's predecessor.[12] Historically speaking, this finding is a bit ironic. One of the earliest paleogeographic connections to be hypothesized and rigorously tested in recent decades was in fact between the east coast of Australia and the west coast of Laurentia—but for Rodinia time. And repeated testing made the likelihood of an Australia-Laurentia connection in Rodinia increasingly low. Thus, although the pair did not seem to be neighbors at the time of Rodinia, they appear to have been at the time of its predecessor. Much like a deck of cards then, the supercontinent cycle reshuffles the global order of things, turning neighbors into strangers and strangers into neighbors.

• • •

Placing Australia in Rodinia's predecessor means that we must look beyond Hoffman's United Plates of America to reconstruct the entirety of the ancient supercontinent. It also thus makes sense to look beyond Hoffman's term "Nuna" for the larger supercontinent. The Inuit term Hoffman had used was a perfect description of the core of the larger supercontinent. In addition to Laurentia and Baltica, which Hoffman had originally referred to as Nuna, this core of central continents included Siberia.[13] All of the mountain belts of these three core continents are old, ranging from about 2 to 1.75 billion years ago—nothing as young as Australia, which didn't ultimately join the supercontinent until as late as 1.6 billion years ago.[14] The formation of the supercontinent's core, assembling nearly 200 million years before the final amalgamation of the complete supercontinent itself, reminded my colleagues of a similar relationship: the formation of Gondwana about 170 million years before Pangea.

Gondwana has always had an awkward position in the supercontinent cycle. It was truly massive: recall from chapter 1 that it was composed of all of today's southern continents. Its sheer size has made it such that some researchers simply refer to it as a supercontinent, possibly without considering the implications of the definition. Others argue that as Gondwana is half of Pangea, albeit the larger half, it nonetheless can't be referred to as a supercontinent because it's just a piece of the larger puzzle. My colleagues and I argue that you can solve both this Gondwana identity crisis as well as the Nuna/Columbia semantic standoff all with one simple proposal.

We realized that every supercontinent gets started with the early assembly of a large building block. With the help of none other than Paul Hoffman, we decided to call this early and large building block a "megacontinent."[15] Pangea had Gondwana form as its most major and early building block, about 170 million years before supercontinent Pangea finally amalgamated. As we've just discussed, the megacontinent of Nuna assembled about 200 million years before supercontinent Columbia finally amalgamated. And while not mentioned directly in chapter 2, because it is the most speculative of the megacontinents, even supercontinent Rodinia is thought to have been potentially preceded by roughly 200 million years by the assembly of megacontinent "Umkondia," named after a major eruption of lavas in South Africa that can be found on the continents surrounding it. Thus, each supercontinent likely had a megacontinent about half its size form some 200 million years earlier. As noted before, Rome wasn't built in a day, and neither is a supercontinent. A megacontinent may be a critical first major step in the larger process.

But if we have taken Hoffman's name for the supercontinent away from him (or at least reassigned Nuna to megacontinent status), we certainly didn't take away his critical speculation about the potential importance of supercontinent Columbia. And I use the word "speculation" not out of disrespect but to acknowledge how many years Hoffman was ahead of his time and also because, in his 1989 paper in *Geology*, that was the exact word he used in his title: "Speculations on Laurentia's First Gigayear (2.0 to 1.0 Ga)" (a gigayear being a stretch of 1 billion years; "Ga" means "gigaannum" for a billion years ago).[16] But before Hoffman speculated, he amassed a wealth of geologic observations indicating that Laurentia's first gigayear was a special time and place in Earth history. In his previous work, Hoffman had already united the plates of America. Now he would investigate what happened as a consequence of those continental plates coming together and forming a supercontinent.

To start, I will let Hoffman speak for himself and cite his scientific paper directly: "Laurentia's first gigayear resembles a symphony in four movements: 2.0–1.8 Ga (allegro), 1.8–1.6 Ga (andante), 1.6–1.3 Ga (adagio), and 1.3–1.0 Ga (allegro)."[17] Hoffman has the ability to be poetic and precise at the same time in his scientific writing. Here he uses the range of tempos in classical music as a remarkably apt metaphor for describing the tempo of the geological "movements" through time. The first two movements, allegro (2–1.8 Ga) and andante (1.8–1.6 Ga), refer to what we just discussed: first, the brisk tempo of assembly of the Nuna megacontinent and, second, the moderately slow tempo of the final amalgamation of the Columbia supercontinent. The third movement, adagio (1.6–1.3 Ga), slowed things down as the newly formed supercontinent matured. (The fourth movement, a return to mountain building, is the tale of the Grenville Orogen discussed in chapter 2, heralding the next supercontinent cycle.) Mozart's final symphony, no. 41, known as "Jupiter," follows closely the standard four movements that Hoffman saw reflected in the geologic record.

It is during the third, slow movement that Hoffman noted the occurrence of some of the most peculiar and debated rocks on Earth: anorthosites. Most everyone reading this book has seen anorthosite—I guarantee it. This is because the white surface of the Moon is essentially entirely anorthosite; the dark regions are the "maria" (Latin for "seas") which are basaltic lava flows. The reflection of the sun's rays off anorthosite is why the Moon is so bright. Anorthosite is nearly entirely composed of the mineral plagioclase, which is also

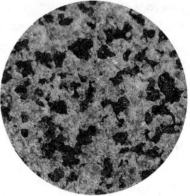

Figure 28. Anorthosite with white calcium-rich plagioclase crystals. Note ruler with one-centimeter ticks on the left. Zoomed-in photo on the right is about 2.5 centimeters across. Photo credit: Ross Mitchell.

common in the other slowly cooled igneous rocks we've mentioned, granites. But unlike granites, in which plagioclase is often rich in the element sodium, in anorthosite, plagioclase is rich in calcium (fig. 28). Being white and translucent in color, calcium-rich plagioclase in anorthosite makes the Moon reflect the sun's light brightly. So what are these Moon-like rocks doing on Earth during the maturing stage of supercontinent Columbia? Laurentia at this time had an abundance of anorthosite never encountered before or after. Why did so many "Moon-like rocks" form in supercontinent Columbia? Hoffman had to wonder, even speculate.

As we've described, the process of subduction on Earth, which takes oceanic crust and pushes it back down into the mantle, is the best recipe for getting water to the base of the crust, and which is also why our having granite is unique to Earth—because we have plate tectonics which recycles water. As the oceanic crust ages, it incorporates increasing amounts of seawater into the pore spaces between minerals. Then, upon being subducted, the slab of oceanic crust will dehydrate as it encounters heat in the depths of the mantle as it descends. Recall the "just add water" recipe for making granites in subduction zones from chapter 2. But the minerals in anorthosite do not store any water in them, implying there was little (or no) water around when they crystallized. Anorthosites therefore are not likely to have formed in response to subduction. So although anorthosites, like granites, are slowly cooled igneous rocks that form large bodies, anorthosites, unlike granites, are not the products of subduction zones. The Moon likely never had subduction or plate tectonics, but its Swiss-cheese-like visage is because it's made of calcium-rich anorthosite.

The other reason anorthosite is peculiar is that the rocks that melt to form its magma originate from such great depths. Oceanic crust, by comparison, forms from melt created at shallow depth within a few kilometers of Earth's surface—think Iceland, where the pulling apart of the plates on either side of the Atlantic Ocean causes the molten mantle to rise up to the surface. Anorthosites, on the other hand, form from melt created at the very base of the crust, near the top of the deep mantle. This interface is the compositional boundary between the more buoyant crust on top and the denser mantle below. For anorthosites to form, no direct connection to plate tectonics need necessarily be invoked. But certainly heat from the hot mantle beneath the plates is a key ingredient. With a heat source focused in the mantle, then all that is required is the compositional boundary at the base of the crust to let melted rock accumulate in a pond before it eventually rises up toward the surface due to thermal convection. Hoffman therefore knew that the profound bloom in anorthosites during Columbia amalgamation might be telling him something about a superheated mantle beneath the supercontinent.[18] But before completing Hoffman's tale, we must have a better understanding of the basic physics that made him look to the underlying mantle for answers.

.  .  .

Geology does not rely exclusively on observational evidence. Theory is as essential to testing a geologic hypothesis as are data. In fact, the tension between data and theory is often a competition that brings out the best in both of these approaches to the scientific method. This tension exists and is helpful because data and theory can be very much independent of each other. Sure, there is some overlap: generating data requires theoretical assumptions in everything from making measurements on an instrument to analyzing the meaning of the new data. Conversely, theoretical models require using parameters or numbers that are often derived from experimental measurements. But, broadly speaking, testing a hypothesis by acquiring new data, or by generating a new model, are independent means of using the scientific method. Back to our scientific crossword puzzle, data and theory should cross-check just like acrosses and downs.

Is it feasible that an amalgamating supercontinent gave rise to the mantle upwelling Hoffman envisioned was created underneath Laurentia and presumably beyond? Hoffman's idea was largely based on geologic data, but a computer model would be needed to make a coherent theory. Modeling could attempt to explicitly test Hoffman's idea that the formation of a supercontinent would heat up the mantle over which it forms.

Not all geologists are rock jocks like Paul Hoffman or Erin Martin with her trusty sledgehammer. Some geologists, often preferring to call themselves "geophysicists," are proud computer nerds. Nan Zhang, a colleague of mine in Beijing, is exactly that. Zhang models convection in the mantle. While this might seem straightforward, it is anything but. Fortunately, computer processing speeds have accelerated unbelievably and continue to do so. This is critical because Zhang is not just modeling simple convection but convection on Earth. The more Earth-like (i.e., realistic) you want your model to be, the more complicated it becomes and the longer the model takes to run. Processing power is such a limitation on Zhang's work that he must rely on supercomputers to complete his simulations.

Many numbers go into the many equations of Zhang's models, such as mantle viscosity. Viscosity, or resistance to flow, can be thought of in terms of water (low viscosity) or honey (high viscosity). Although the mantle is stiffer (more viscous) than either of these fluids, it nonetheless also flows. The tricky thing about a parameter like mantle viscosity is that it is difficult to determine. We certainly have means of detecting how stiff the mantle is, but there are large uncertainties associated with those estimates, and things get even more uncertain the deeper one goes in the mantle.

The relatively shallow mantle is quite well constrained by measuring the rate at which the mantle has rebounded in Scandinavia after its ice sheet melted following the last glacial maximum. Once I got to attend a conference in Lulea, Sweden, which is the bull's-eye center of where the rebound of the ground is fastest, occurring at a rate so fast that the town's docks have to be constantly rebuilt to lower ground as the earth bounces back. But this rebound only tells us about the stiffness of the shallow mantle. The deep mantle is thought to be much stiffer, and a more precise estimate of its viscosity is challenging. Because of this uncertainty, Zhang must explore a range of possibilities, seeing how sensitive his model results are to changing mantle viscosity by a factor of 10 or more. The good news is that as Zhang's computer simulations become more Earth-like due to increasing processing power, they are also becoming more realistic as measurements of parameters like mantle viscosity are made.

But the power of modeling does not merely come from making Earth-like models. In fact, sometimes just the opposite can be just as helpful. David Bercovici of Yale University, a card-carrying modeler like Zhang, likes to joke that, when it comes to modeling, "reality is overrated." Recall how large the uncertainties in the variables that get put into these models can be, or the limitations that computer processing power can impose on the "reality" of the model. Bercovici's comment implies that it may be naive to think of these models as *ever* resembling reality. He really means that one can learn the potential capabilities of a system by twisting all the knobs and exploring all imaginable, feasible possibilities.

•    •    •

Zhang's computational contributions to understanding the supercontinent cycle started when he was a graduate student working with his mentor, Shijie Zhong. Zhang and Zhong had just the model to test Hoffman's speculation that supercontinent formation could form a mantle upwelling beneath it. But it's also important to understand why their particular model was particularly significant by the time it came to be. Before the work of Zhang and Zhong, a supercontinent was thought to heat up the mantle beneath it through a "thermal blanket" effect. Continental crust, compared to oceanic crust, is thick and thus slower to let mantle heat escape. The idea of the thermal blanket was that if all the continental crust was aggregated in one place (a supercontinent), then the heat flow would become reduced there, allowing the mantle beneath the supercontinent to heat up over time.[19] Simple ideas are great, but they can also be wrong, or at least insufficient for the problems they purport to solve. In the years since Hoffman's speculation, modelers had made calculations that indeed confirmed the existence of the thermal blanket effect but also questioned its importance relative to other processes. It was estimated that the thermal insulation of a supercontinent could only account for about a 20°C increase in temperature, which would barely make it much different from the temperatures of the mantle under the oceanic crust.[20] Another mechanism was needed if indeed a supercontinent was capable of creating a hot mantle upwelling.

In 2007, Zhang and Zhong came through. Theirs was one of the first models of mantle convection to be conducted in 3-D. About a decade earlier, when Zhong himself was a graduate student at Caltech working with Mike Gurnis, they were pioneering the first mantle convection models to inform why supercontinents assemble.

But these models were only done in 2-D. Because Earth is spherical, the 2-D Gurnis and Zhong models were not Earth-like in this respect, but they still were useful for making important discoveries. In the first chapter on Pangea, we accepted the basic physics that continents would tend to drift "downhill," which would predict that a supercontinent would amalgamate over a topographic low where the underlying mantle was downwelling. Part of the importance of modeling is demonstrating that common sense actually continues to make sense when you test it with a complex model that's as realistic as possible. In *Nature* in 1988, Gurnis had done exactly that, demonstrating that, as one might expect, a supercontinent will tend to form over the topographic valleys where there's a pronounced downdraft in mantle convection—essentially confirming the commonsense notion that a continent would settle over a strong mantle downwelling, much like a rubber ducky floats over to the downdraft of a draining bathtub (fig. 29).[21] During his years as a graduate student, Zhong would work with Gurnis to make this model increasingly sophisticated, enabling them to test new dynamics of supercontinent assembly.

Fast-forward to Zhong as a faculty member at the University of Colorado, Boulder and his then promising student, Zhang, and they were able to take a new leap forward by running their simulations in three dimensions. But they also had another trick up their sleeves. In addition to the computational efficiency allowing them to make a 3-D mantle, they also had a new idea to test. The concept would be to explore what would happen to mantle convection *after* the supercontinent formed. Building off Gurnis's influential work, they could already assume that the supercontinent would form over a mantle downwelling. For a next step, then, they could just drop a supercontinent on top of the cold downwelling and sit back and watch to see what would happen.

First their model would start in chaos. The team of Zhang and Zhong had to first prove that a massive mantle downwelling—a

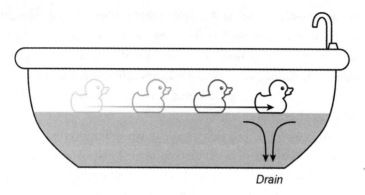

Figure 29. A continent drifting over the convecting mantle will move toward a cold mantle downwelling, much like a rubber ducky floats toward the drain where the water is getting sucked down.

superdownwelling—would even develop in the first place. So they started with a random array of downdrafts and updrafts in the mantle—many of each. Then, they pressed start and watched the model evolve. Over time, what they found is that the downwellings and upwellings in the mantle would each begin to become more organized by combining, thereby steadily reducing the number of each. Downwellings merged with downwellings, upwellings with upwellings, until eventually there was just one of each.[22] What does that look like? In one hemisphere, convection in the mantle is sinking (downwelling), and in the other hemisphere, the mantle is rising (upwelling). In the terminology of Zhang and Zhong, this pattern reflects "degree 1" convection because there is only one convection cell (fig. 30). We can think of degree 1 convection as "one-sided" convection, like an oven with only one heating element at the bottom of the oven, where the sinking cold air is replaced by rising warm air.

Once the duo had created their degree 1 flow—forged from a chaotic beginning—they took the next step to see how the pattern of convection might change if a supercontinent were to form. They assumed that the continents would collect over the hemisphere of mantle downwelling, following our metaphor of taking the plug out of a child's bathtub and watching that rubber ducky glide right over to where the drain is sucking the water down (fig. 29). Zhong and Zhang took a continent about the size of Pangea and plopped it squarely in the center of the superdownwelling and watched to see how the underlying mantle would react. The degree 1, one-sided heating pattern didn't last long.

What they found was that mantle downwelling started to occur all around the edge of the supercontinent. This phenomenon was due to subduction. Because seafloor spreading creates new oceanic crust all the time, subduction must also be destroying oceanic crust in equal proportions somewhere in order to preserve Earth's constant surface area. As the continents were drifting toward the downwelling, they were consuming the intervening oceanic crust by subduction as they did so. But then, once the new supercontinent forms, all those subduction zones once between the continents cease to be. Subduction must therefore start up somewhere else. And it does so all around the edges of the newly formed supercontinent (fig. 30).

What Zhong and Zhang found was that circum-supercontinent subduction rather quickly transformed the superdownwelling over which the supercontinent had formed. Remember the principle of convection is that what goes down, must come up. The ring of subduction around the supercontinent thus caused the mantle caught between all this concentrated downwelling to react. This reaction was a "return flow" which meant that the downwelling once beneath the supercontinent evolved to become an upwelling.[23] Now Earth's convection pattern was different. Their experiment started with one mantle upwelling and an opposite mantle downwelling with a new supercontinent formed on top of it. But the presence of that new supercontinent, and the ring of subduction it generated around it, had transformed that downwelling into an upwelling. Now Earth, according to their model at least, had acquired a convection system similar to that of modern Earth, with two upwellings mirroring each other and bisected by a ring of mantle downwelling in between (fig. 13).

This configuration is referred to as "degree 2" flow due to the addition of a second mantle upwelling forming beneath the supercontinent (fig. 30). One-sided heating (degree 1 convection) had turned into "two-sided" heating (degree 2 convection), now like an oven with heating coils at both the top and the bottom. Zhong and Zhang had done it. They had tested and proved Hoffman's speculation: form a supercontinent and the mantle downwelling over which it formed will transform into an upwelling. The supercontinent-induced mantle upwelling could explain the bizarrely abundant occurrences of deep-seated melts from the mantle forming a bloom of anorthosites

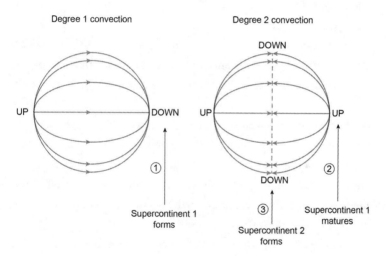

Figure 30. Mantle convection in 3-D with upwelling and downwelling. Random convection evolves naturally to degree 1 convection (*left*), that is, one hemisphere of upwelling and one hemisphere of downwelling. Step 1, forming a supercontinent on the degree 1 downwelling (toward which the continents would flow "downhill" and collect above; fig. 29) converts the downwelling into an upwelling by return flow, thus creating in Step 2 degree 2 convection (*right*), that is, with two upwellings opposite of each other. Degree 1 and degree 2 convection can be thought of as "one-sided" or "two-sided" heating, like an oven with one heating coil on the bottom of the oven (degree 1) or with two heating coils, one each at the top and bottom of the oven (degree 2). Formation of the next supercontinent, Step 3, will occur then in a world of degree 2 convection, which will be discussed in chapter 5.

across much of Laurentia. Only the formation of a supercontinent could explain the development of the two-sided heating observed in Earth's mantle today.

•   •   •

In the final section of his paper, Hoffman had snuck in another speculation that went even further. Under the notion that proposing controversial ideas in terms of a question makes them more palatable to skeptics, Hoffman titled this final section of his paper with a query, "Earth's First Supercontinent?" Once again, he had very good reason to warrant speculation. There had been younger supercontinents than Columbia, namely Rodinia and Pangea, but neither of these, particularly Pangea, had abundant anorthosites to speak of. Pangea certainly had evidence for having created a mantle upwelling beneath it (fig. 13). In fact, due to this thermal legacy from only 200 million years ago residing to this day in the lower mantle, the evidence for mantle upwelling beneath Pangea is the strongest case for supercontinents influencing mantle convection. But even with this superheated mantle beneath it, Pangea experienced very few intrusions of anorthositic melt rising from the base of the crust. What made the mantle upwellings beneath the older supercontinents special in this regard, Hoffman wondered?

The answer was in the question. They were older. Earth was born in fire, and it has been cooling down ever since. One of the most fundamental questions in geology is how quickly Earth has cooled.

But whether fossil heat (leftover from the planet's formation) or radiogenic heat (created by the decay of radioactive elements that continue to produce heat, like uranium and potassium), their effects over time are strikingly similar: both imply that Earth used to be a lot hotter. The loss of fossil heat makes immediate sense. Even if the rate of heat loss of fossil heat has been constant over time, this would imply that Earth used to be hotter. The pace at which radiogenic heat is lost, however, is notably not constant. Remember that each radioactive element has a half-life, or a fixed amount of time over which half of its radioactivity decays away. Because of exponential decay, radiogenic heating of the mantle by elements like uranium, potassium, and thorium would have been quite significant early in Earth history, but increasingly less so over the course of time. Putting these two effects together—less and less fossil heat and less and less radiogenic heat—means Earth was much hotter in the past than it is today.

This long-term change, Hoffman realized, could maybe explain why Columbia and Rodinia, the older supercontinents, had particularly special mantle upwellings that formed anorthosites. Supercontinents will always tend to heat the underlying mantle, as Zhong and Zhang showed, by turning the mantle downwelling over which a supercontinent forms into an upwelling. But if the average temperature of the mantle happened to be hotter in more ancient times, it follows that the mantle upwellings that formed under the older supercontinents, Rodinia and particularly Columbia, would have been hotter than the upwelling under Pangea. The first time in Earth history that widespread anorthosites had formed had to be special for some very good reason, Hoffman reasoned. In addition to the heat from the supercontinent-induced mantle upwelling, the on-average hotter ancient mantle provided a little extra heat flow at the base of the crust to make the peculiar melts that form anorthosites. By this logic, Hoffman speculated that Columbia might very well have been Earth's first supercontinent.[24]

In the next chapter, going back in time even deeper still, we will do our best to challenge this assertion. But we will also encounter some of the very good reasons for why the amalgamation of Columbia might have been the birth of the supercontinent cycle and the first of its kind.

# 4

# THE UNKNOWN ARCHEAN

At present, our knowledge . . . is both frustrating and exhilarating—
frustrating because we are certain of so little, but exhilarating because
we know anything at all . . . the companion of ignorance is opportunity.

ANDREW H. KNOLL, *Life on a Young Planet*

Has Earth always made supercontinents? Or is there a point during
our adventure farther back in time when we no longer find compelling evidence for an older supercontinent? If so, are there compelling
reasons for why the young Earth might not have created supercontinents? For the final part of our journey back through time—before
we finally cast off into the future in the last two chapters—we will
explore the mysterious recesses of Archean time.

Earth was essentially a different planet in the Archean. In the geologic timescale, the longest subdivisions of time, like acts in a play,
are eons (fig. 27). The eon of Pangea is called the Phanerozoic, from
the Greek *phaneros*, "visible, evident," and *zoion*, "animal," inspired
by its fossil record, which by then included animals large enough to
see with the naked eye. The Phanerozoic began 541 million years ago
and continues to the present day. The next oldest eon, the Proterozoic, began 2.5 billion years ago, hosted both Rodinia and Columbia,
and was named from the Greek *proteros*, "former," and *zoe*, "life," for
the first appearance of multicellular, animal life, albeit in the form

of microfossils seen only under a microscope. The third oldest eon, the Archean, which began 4 billion years ago, is named after only one Greek root, *arkhaios*, "ancient." There's no mention of nascent life-forms. No mention of anything other than its being very old. The only older eon, the Hadean, named for the Greek god of the underworld, is given such an ominous name because it is defined by the time before which we don't even have a single rock preserved. So looking on the bright side, the Archean may be ancient, but at least it had rocks that we can map and sample to gather evidence of our primitive beginnings.

But Archean rocks are different. Even though continental crust formed during the Archean, Archean geologists don't talk about continents forming. Instead, they refer to the rafts of continental crust in Archean time as "cratons," mentioned briefly earlier, from the Greek *kratos*, "strength." Cratons can be thought of as the ancient nuclei of younger continents. Every continent on Earth today contains a handful of Archean cratons. On average, each Proterozoic continent contains about four Archean cratons (fig. 31). So cratons are small, but given their name, they are also mighty. On Earth's dynamic planet, they have survived for more than 2.5 billion years.

The most defining aspects of cratons are that they are obviously ancient and long-lived, which means they are resilient for some reason. But why? And what does continental crust being resilient even mean? Not all continental crust is built to last forever. Continental crust rich in low-density granite is buoyant and so resists subduction back into the mantle. But even continental crust can be eroded over time. Continental crust can certainly be eroded by wind and weather, but it can also be eroded by subduction. Dense subducted lithosphere acts like a sinking anchor, dragging downward anything that is attached to it. Thus, subduction slowly but steadily drags more and more crust back into the mantle. But unlike younger continental crust, ancient Archean cratons are forever.

Cratons are long-lived and resilient because they are strong. They

Figure 31. The number of Archean cratons (with the names of each) embedded within each Proterozoic continent (thick outlines). The regions outside of the continent outlines represent relatively young crust added in the last few hundreds of millions of years.

are strong because they are thick. And their thickness isn't just in their crust but also in their underlying lithosphere. Even though geologists separate the crust from the mantle in terms of their different chemical compositions, the tectonic plates that move around are composed of crust as well as a decent chunk of underlying mantle. This bit of mantle underlying the crust of each plate is called the "mantle lithosphere," and this hard layer forms the stiff base of the tectonic plate that glides over the underlying weak layer called the "asthenosphere." It is the low resistance to flow of the asthenosphere that allows the tectonic plates of the lithosphere to move horizontally, and its difference from the lithosphere is more physical than compositional.

Even though every tectonic plate has a good chunk of mantle lithosphere underlying it, cratons have the most. In fact, the mantle lithosphere underlying Archean cratons is so thick—hundreds of kilometers—that the crusts of cratons are often said to have a keel, like a ship.[1] And just like the keel running down the hull of a ship, the cratonic keel also provides stability as the plate "floats" above the underlying soft and convecting mantle. As a general rule, the thicker the cratonic keel, the more stable and more likely to survive the next supercontinent cycle it is. Oceanic plates, even though they also have thick lithospheric mantle underneath them, are thinner than continental plates and, because they are denser, preferentially get subducted when plates converge. Thin continental plates too are not long for this Earth: think of the thinned crust of the southwest United States, which is highly tectonically active and heavily faulted, and home to the famous, or infamous, San Andreas Fault.

But cratons don't have such faults, and it's not inaccurate to think of them as essentially flawless. Because of their strength, cratons have survived for billions of years and most of them don't show any signs of stopping anytime soon. Of Earth's thirty or so cratons, only a small handful have shown any signs of weakness. The thick keels of mantle lithosphere underneath the Wyoming and North China cratons, after about 2.5 billion years of stability, have recently come under threat. Both of these cratons are positioned along the Pacific ring of fire, and the past few hundreds of millions of years of subduction of the Pacific plate under these two cratons is finally taking its toll. Subduction can, bit by bit, erode the overlying plate as the down-going plate pulls these fragments down with it. Thus, unlike most cratons that still retain their thick underlying keels, those underneath Wyoming and North China have been considerably thinned. Even then, the ancient crust of these two besieged cratons remains safe. Although their mantle keels are much thinner, they were originally so thick that what remains is still adequate to protect the overlying crust. Above all, it is the rarity of these exceptions that proves the rule that cratons are typically stable for life and resist being modified by even billions of years of subsequent tectonic forces.

●   ●   ●

Cratons aren't only endowed with great thickness, they are also well endowed with mineral resources. The Archean was not only a golden age of cratonization. It was literally a golden age. Over half of the world's known gold deposits hail from Archean cratons.[2] What was the secret to cratons forging gold? If I knew the secret, I probably wouldn't be writing this book! But we certainly have good ideas about the critical ingredients that contributed to the recipe of concentrating gold in Archean rocks and why such deposits are relatively rare in younger rocks.

Archean cratons are known for their "granite-greenstone" belts. We've covered granites at this point. Greenstones refer to basaltic rocks (recall that basalts are primitive rocks derived directly from the mantle) that are very similar, or arguably identical to, oceanic crust. In some cases, basalt occurs as lava flows between layers of sedimentary rock. But in most cases, the greenstones comprise a large block or region of exclusively lava flows, similar to the upper layers

Figure 32. Aptly named "pillow" lavas that erupted underwater in New Zealand, with geologist for scale. Photo from https://www.geotrips.org.nz/trip.html?id=73. Courtesy of Julian Thomson, Out There Learning Ltd, NZ.

of oceanic crust. And often the style of lava flows, known as "pillow lavas" for their bulbous shape (fig. 32), signify that they were indeed from lava erupting underwater like pillow lavas in modern oceanic crust. Archean granites often get most of the attention because it is their impressive volume and compositional buoyancy that caused cratons to grow in size and, in some cases, to be built up high enough to emerge above sea level billions of years ago—much like modern continents. But it's the greenstones that are endowed with the wealth of Archean gold.

Greenstones, unlike their surrounding granites, contain abundant metals, one of which is gold, but they are also endowed with economically important ones, including nickel and copper. Thus, greenstone can be thought of as the "source bed" for Archean gold.[3] But how does gold get so concentrated, and how can it be readily extracted so that gold mining is profitable? This is where another peculiar feature of Archean Earth might have come into play: komatiites. Named after the Komati River in southern Africa, komatiites are a spectacular-looking volcanic rock. With long, bladed, and crisscrossing olivine crystals, komatiites exhibit a stunning texture described as "spinifex," named after the spiny bush in the Australian outback that pricks and pokes.

But the most important thing to know about komatiites is that they are remarkably rich in magnesium, and this indicates that they represent the hottest lavas (possibly as high as 1,600°C) ever produced on Earth.[4] Nearly all known cases of komatiite are more than 2.6 billion years old, that is, they are nearly all Archean in age.[5] They just don't make 'em like they used to: by the end of Archean time, Earth had cooled down too much to make high-temperature komatiite melts.

Thus, although still unproven, one theory is that superhot komatiite melts were critical in concentrating gold scavenged by greenstones.[6] Unlike younger, lower temperature melts, komatiite melts might have had a greater capacity to dissolve sulfur—and with it gold—during their ascent to Earth's surface. It might have also been important that komatiites were formed quite deep, as deep as the upper mantle. Therefore, in their transit from the upper mantle through the entire crust, komatiite melts were able to liberate and concentrate more and more gold as they percolated up. The temporal restriction of komatiites to the Archean may therefore explain the Archean golden age.

• • •

More riches than gold resulted from the Archean eon of cratons. More than half of the oxygen that we breathe in the atmosphere today arguably came about immediately after craton formation. Earth today certainly has a higher concentration of atmospheric oxygen (~22%) than it did 2 billion years ago, but the largest increase in oxygen rose right after the end of the Archean in the well-named "great oxidation event" or "great oxygenation event."[7] There are of course many theories for what might have caused this singular event. As a supercontinent researcher, my favorite relates to the rise of continents.

Photosynthesis is the best source for accumulating oxygen in the atmosphere. The simple recipe of organisms taking in carbon dioxide, water, and sunlight to breathe out oxygen never disappoints. But if stromatolites formed by photosynthetic bacterial communities are as much as 3 billion or perhaps 3.7 billion years old, why did it take so long for the great oxygenation event to occur?[8] Why didn't the great oxygenation event happen earlier? An often forgotten part of the photosynthesis equation is that, along with sunlight, nutrients are required.[9] In the modern world, if you don't provide your plants with the chemical nutrients they need from sufficiently nutrient-rich soil, the plants won't take and photosynthesis fails—the same would have gone for photosynthetic bacteria in the Archean ocean. They needed nutrients. This is where rocks start to become important.

Although nitrogen is the most important nutrient in the ocean on human timescales, on long geological timescales, the most important source of nutrients is the element phosphorus.[10] This is because phosphorus, an essential part of DNA, is a limiting nutrient. Its source lies in the crust; specifically, phosphorus is concentrated in a mineral called apatite. Apatite is most common in igneous rocks where it is known as an "accessory mineral," essentially a mineral that is invariably present, but only in small amounts. It is the weathering of igneous rocks that plucks phosphorus out of the apatite crystal structure. So you could argue that the stability of the apatite is a rate-determining-step for the proliferation of life! Although rocks can be weathered underwater, it's not really as efficient as the weathering that occurs in rocks exposed to the weather. Thus, the rise of continents above sea level was pivotal in the rise of phosphorus, which in turn was critical to the increase in oxygen.[11] Finally exposed to efficient weathering, the cratons started to shed phosphorus housed in apatite crystals. Critical nutrients thus started to flood into the oceans, allowing photosynthetic bacteria to start living up to their full potential for the first time.

We can now perhaps explain why photosynthesis, long since evolved, waited until about 2.4 billion years ago to finally show what it could do. It had previously been held back by limited nutrient supplies. In this way, the story of life, the oceans, and the atmosphere is intimately related with the coevolution of the rock record. But concerning our story about plate tectonics, the real question then becomes what this early rise of the continents meant for the possibility of an even more ancient supercontinent.

• • •

There are two fundamental questions that raise doubts about the existence of an Archean supercontinent. First, was there even enough continental crust available at that time to form a large supercontinent? Second, was plate tectonics fully operational to assemble most of the continents in a supercontinent? Both of these questions invite great debate. The first question concerning the growth of Earth's crust goes back decades. The second debate concerning the antiquity of plate tectonics is surprisingly recent, reaching a fever pitch about a decade ago and still raging on.

How has Earth's crust grown over time? Taking Earth at face value—literally, just at its surface—the amount of its continental crust has increased over time. That is, if one simply plots the volume of the continental crust preserved at Earth's surface, it increases progressively over time.[12] This systematic increase in the number of rocks preserved at Earth's surface would appear to imply crustal growth. But the problem is more complicated than just what evidence we have at Earth's surface. It's quite possible that some decent amount of crust created at Earth's surface has been subsequently recycled back into the mantle via subduction, called crustal recycling. Geochemical processes suggest that crustal recycling was all the rage back in Earth's early days, even more common than it is today.[13] So the lack of ancient preserved crust could simply be an artifact of more vigorous ancient crustal recycling compared to modern times.

In addition to this caveat concerning crustal recycling is another issue: "crustal reworking." How does this differ from crustal recycling? Most of Earth's crust is created by melting the mantle. Melt the mantle and the first result is basaltic crust above it; then melt that basalt and the next result is the chemically buoyant, granite-based continental crust. But once Earth has continental crust, there is more than the mantle to melt. Crust can also be melted, if it gets hot or wet enough. Melt crust and you get, well, more of the same . . . more crust. But this isn't a net increase in crust, it is merely crustal reworking. Therefore, some amount of the preserved crust on Earth's surface shouldn't be considered as part of the net sum of crustal growth.[14] Some appreciable amount of Earth's present continental crust has been formed by melting more ancient continental crust—continental cannibalism, if you will. Combine the effects of crustal recycling and crustal reworking and the simple story that the distribution of surface ages tells us of apparent crustal growth over time becomes less convincing. The amount of crust Earth has formed over time will be underestimated (due to crustal recycling) and biased toward recent times (due to crustal reworking) if one takes Earth's preserved surface at face value.

Even if the preserved surface rocks give us a biased picture of Earth history of crustal growth, these rocks tell us how much long-lasting continental crust was available at any given time to construct a supercontinent. Furthermore, when continents collide and mountains are built up between them, the continental crust becomes thickened and therefore less likely to ever be subsequently subducted.

Continental lithosphere floats on the underlying convecting mantle like an iceberg floats in the ocean: if it seems big above sea level, there's an even deeper root lurking beneath—hence the common saying "the tip of the iceberg"! This is the principle of "isostasy" (Archimedes's "Eureka!" moment of volumetric displacement), and it implies that tall mountains have even more impressively deep roots. The roots of the thick and deep orogens that sutured continents together into Earth's first supercontinent are therefore still preserved today.

Earth's percentage of preserved crust was essentially nil until 3 billion years ago, and it had reached as little as 20% of its present value by 2 billion years ago. Thus, as the supercontinent Columbia began to take shape, Earth's growing crust was beginning to pick up the pace, making the formation of a supercontinent feasible from the perspective of crustal growth (fig. 33). Before Columbia, it is possible that Earth either hadn't grown, or at least hadn't preserved, enough continental material to construct a large supercontinent.[15] As there was less continental crust at Earth's surface in the Archean, then the cratons would have been submerged in even more of a water-world compared to the continents today.[16] Clearly Earth could make cratons in the Archean, but cratons pale in comparison to the size of continents, where, recall, each continent today contains about four cratons on average (fig. 31). One could argue that, for some reason, cratons broke up into smaller pieces than continents—but that reason may also be related to yet another reason why an Archean supercontinent may have been unlikely.

It is not only crustal growth that challenges the plausibility of an Archean supercontinent. A fundamental question is whether plate tectonics was yet fully operational. Just because Earth has a global plate tectonic network today doesn't mean it always did. Look at Mars or Venus. Some scientists have speculated that both these planets might have had plate tectonics sometime in their distant pasts.[17] But they certainly don't have plate tectonics anymore. Plate tecton-

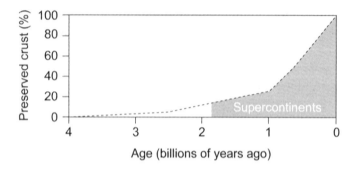

Figure 33. Crustal growth and supercontinents. The gray region shows when super-continents formed (assuming Columbia is the oldest) and the amount of crust on the surface of the earth. Scientists believe that the Earth did not have enough crust to form supercontinents until around 2 billion years ago. Adapted from A. M. Goodwin, *Principles of Precambrian Geology* (London: Elsevier, 1996).

ics can't be taken for granted, even on Earth, and especially in Earth's distant past. We tend to forget that the theory of plate tectonics was developed almost entirely by trying to explain Earth's recent or even current plate boundaries. Whether this theory applies to Earth for all time is a matter of ongoing debate.

Today's geologists generally favor one of two philosophical approaches to the possibility of ancient plate tectonics. First, there are the uniformitarianists. Recall this was geologist pioneer Charles Lyell's famous maxim: "the present is the key to the past." The others are fans of that other scientific giant named Charles, namely, Charles Darwin. Evolution as applied to geology says processes in the past evolved to what they are today.

If we have plate tectonics today, uniformitarianists contend, the simplest interpretation is that we had it in the deep past. They ask Precambrian geologists if they can find evidence of the equivalent of today's plate tectonics in the rock record.[18] The evolutionists don't believe we need plate tectonics to form the oldest rocks.[19]

•　•　•

So what are Earth's oldest rocks and what can they tell us? The setting for the landmark discovery was the remote lakes and tundra of northern Canada. And of course, Paul Hoffman is somehow, however slightly, involved. Sam Bowring, a PhD student at the University of Kansas at the time, was wise enough to team up with Hoffman working on Great Bear Lake in the Northwest Territories. Bowring would go on to become one of the most important geochronologists in history, which is certainly saying something in a field founded by Arthur Holmes and that included Claire Peterson, the first person to date the 4.543-billion-year-old age of the Earth. By the end of his career, at MIT of all places, Bowring's world-renowned geochronology laboratory took up an entire floor of the Earth, Planetary, and Atmospheric Studies building—which again is saying a lot at a place where every professor is a rock star.

When Bowring got started, ages in millions of years had uncertainties in millions of years; by the end of Bowring's career, ages of millions of years had uncertainties in thousands of years. Bowring always remained a well-rounded geologist even as he honed his expertise in geochronology. He was famous for saying "no dates, no rates." Bowring didn't just want to assign ages to rocks (the "dates"); he wanted to use rocks to compare the "rates" of biologic evolution of the Cambrian explosion to "rates" of extinction during the "Great Dying" of the Permian-Triassic mass extinction.

But at the time of the story we're telling, Bowring was still junior. Bowring had learned from Hoffman a skill that has unfortunately gone out of vogue in geology schooling: geologic mapping. Every question in geology starts with looking at a geologic map, so it helps if students learn how to make one (fig. 34). The process usually starts when a geologist hikes up to a good vantage point overlooking the map area to identify pertinent outcrops—outcrops that are the windows into which one can peer into the structure of the rocks underground, often covered by dirt and forest. At each outcrop, the geologist identifies the rock (igneous, sedimentary, or metamorphic,

Figure 34. Geologic map made by Paul Hoffman of the Slave craton, Northwest Territories, Canada. Colored pencil is used to designate each different rock layer. The oldest layers are found at the southern side of the island, and the layers get progressively younger to the north. Photo credit: Ross Mitchell.

and specifically which types) and then each rock type gets a unique color on the map. Technicolor geologic maps make for mesmerizing wall art. The boundaries between the different colors are also critical: boundaries between different rock types indicate which rock is older and which is younger and if any tectonic motion, such as a fault, is involved. Thus, even before precisely dating any rock at all with geochronology, one can infer the *relative* ages of the rocks in a given field area.

And this is exactly what young Bowring and members of the Geological Survey of Canada did up in the northern tundra of Canada: mapping. The area Bowring was working, the Slave craton, home to Hoffman's PhD research, was known to have some of the oldest rocks in North America. Furthermore, the corner of the Slave craton that Bowring was mapping had some of the oldest rocks in the craton. Bowring was a well-trained student and he followed mapping protocol. Before dating a single rock sample back at the lab, Bowring had identified the oldest rocks in the area. And given the continental and regional context, he already had a sneaking suspicion the sample bags he loaded onto the helicopter might hold some truly ancient secrets.

When Bowring dated his ancient rocks, he found that he had indeed discovered Earth's oldest rocks. Dated by Bowring in 1989, the Acasta Gneiss still remains Earth's oldest known rock to this day, coming in at 4.03 billion years old—only 513 million years younger than the age of Earth itself![20] The Archean, the second oldest eon in the geologic timescale (fig. 27), is set at a round number of 4 billion years ago, signified by the start of Earth's rock record with the Acasta Gneiss.

•   •   •

But as odd as it may sound, the geologic record doesn't in fact start with Earth's oldest rocks. We have something even older. In the sunburned outback of western Australia, in the 1980s, geologists in Australia found some ancient sandstones that contain Earth's oldest terrestrial materials. In addition to the grains of sand dominantly composed of quartz, there were ancient sand grains of zircon. The sandstones were deposited about 3 billion years ago. But the grains were derived from the erosion of still older rocks. Recall how chemically resilient these zircon time capsules are. Over the next few decades, older and older zircon grains would be found in the Jack Hills region of the Yilgarn craton.[21] The oldest Jack Hills zircon dated is 4,404 million years old—a staggering mere 139 million years after Earth's birthdate. Those zircon grains had to be part of some older crustal rocks that formed 4,404 million years ago. A fair share of the Jack Hills zircons are older than the Acasta Gneiss and therefore older than the Archean eon—thus Hadean in age, from Earth's oldest eon.

Other Hadean zircons have been discovered elsewhere in the

world, such as in the Barberton greenstone of South Africa, but the Jack Hills zircons still represent about 95% of all known Hadean zircons, making this region of the Australian outback so critical for understanding the earliest Earth.[22] Three billion years ago, this ancient igneous rock was exposed at Earth's surface, weathered, and shed its zircons that became deposited as a sandstone. This makes the Jack Hills zircons detrital zircons, for they are not found in the igneous rock in which they were forged but as detritus in a sedimentary rock formed at some later date. This by no means diminishes their singular importance. But it does mean that there is great debate over the source rock of the all-important Jack Hills zircons.

Since their discovery three decades ago, pretty much every type of igneous rock has been proposed as the source for the Jack Hills zircons. And the debate shows no signs of abating, but this doesn't mean that we aren't making progress. The stakes are high because if you figure out the source rock of these ancient grains, then you are defining the type of rock Earth first formed—even though it doesn't exist anymore—which paints a picture of the geology of earliest Earth and the potential conditions for fostering the origin of life. From a purely theoretical point of view—considering motive in this investigative case before considering the evidence—one might suspect the Jack Hills zircons came from a granite. Zircons are abundant in granites, but very sparse or nearly absent entirely in basalts. This is because of fractional crystallization, the same process that explains why granites eventually arise naturally out of evolved magma chambers that once generated basalts. Zircon combines the element zirconium with oxygen, and zirconium is said to be an "incompatible element." This means that the positively charged ion, or cation, of zirconium isn't quite the right size or charge to fit into common crystal structures. More suitable, or more compatible, elements like magnesium or nickel will link up to form minerals first. Then, over time, as the magma crystallizes, it has fewer compatible elements to favorably select from, and elements like zirconium finally get selected and zircons begin to crystallize in the evolved, or fractionated, magma chamber. Statistically, it is reasonable to suppose that the Jack Hills zircons were sourced from a granite (in which zircons were plenty), instead of a less fractionated, more primitive source rock like a basalt (in which zircons are scarce).[23]

While a granite source seems logical, the implications it would carry, if supported by evidence, are profound. To follow the old Carl Sagan saying: "extraordinary ideas require extraordinary evidence." Granites forming as early as Hadean time would be quite extraordinary. Granites, as we've discussed with our "just add water" recipe for them, are signatures of both water and plate tectonic subduction systems that dehydrate descending slabs to infuse water into the magmas that most efficiently generate granites. Thus, saying the Jack Hills zircons came from granites could imply quite profoundly that Earth in its earliest days was almost identical to the one we know and love today. Such an interpretation would imply the presence of both abundant water and operational plate tectonics.[24] Given how dependent modern forms of life are on surface water and nutrient cycling due to plate tectonics (recall phosphorus as fertilizer), such modern-like conditions would imply, quite encouragingly, that early Earth was a very hospitable place for life. Thus both these conditions for creating granites are possible, but since they have such weighty implications for habitability so early in Earth history, skeptical scientists must naturally question such a view of Hadean time. The verdict is still out on the true significance of Earth's oldest terrestrial record, the enigmatic Jack Hills zircons.

•   •   •

Even if Earth had fulfilled the prerequisite of plate tectonics quite early in its Archean or even Hadean history, that still doesn't fully address our main concern—whether or not Archean cratons had been clustered together in an even older supercontinent. Operational plate tectonics is certainly necessary for making a super-continent, but it's not sufficient. As we've already discussed, there would need to have been enough continental crust. That criterion, you recall, already had us suspicious of an Archean supercontinent. Plate tectonics also would have had to have been more than merely locally operational to have forged a supercontinent: it would have had to have been global. Ancient subduction could have occurred locally, but such localized tectonic phenomena occur even on the icy shell of Jupiter's moon Europa. Even though Europa has localized subduction, no one would argue that it has plate tectonics.

Plate tectonics, as it is defined for Earth at present day, is a global network of interconnected plate boundaries. How long might it have taken for localized tectonics features to have linked up in a truly global network is very much related to when Earth could have formed its first supercontinent. With one of my colleagues at the Chinese Academy of Sciences in Beijing, Bo Wan, we have shown that subduction structures discovered on a large handful of different cratons were linked to each other about 2 billion years ago in the lead-up to supercontinent Columbia. As of now at least, this is the oldest evidence for a global plate network and, as such, supports Hoffman's speculation that Columbia was Earth's first supercontinent.[25] But no doubt others will follow up our work and endeavor to see if older evidence for a global plate network might be found and thus have possibly allowed yet an older supercontinent to have formed. Although the cards appear to stack up against an Archean supercontinent, it is nonetheless a difficult hypothesis to reject.

What would the alternative hypothesis to an Archean supercontinent be? Before he shifted his focus from supercontinents to Snowball Earth, Paul Hoffman had noted that the granite-greenstone architectures of Archean cratons were truncated at the margins of each craton. That is, it appeared as though the granite-greenstone belts should have continued across a greater area before they were truncated by younger Proterozoic rift margins. This general juxtaposition of old Archean cratons with young Proterozoic rifted margins suggested to Hoffman that most of the cratons of the world had originated as part of some ancestral landmass or landmasses. Even though Hoffman himself left the Geological Survey of Canada for Harvard University, this yearning to understand these most ancient ancestral lands lived on in Ottawa as Dutchman Wouter Bleeker picked up the mantle in the years following. Bleeker is a special figure to me personally as he was my unofficial PhD advisor during my fieldwork up north. Of course David Evans gave me world-class supervision back at Yale and in the laboratory, but during my fieldwork in the remote tundra of Slave craton, Evans trusted Bleeker to take care of me and my development as a field geologist. But before Bleeker and I met, he had written a critical paper that put a new spin on the paleogeographic possibilities of Archean time.

One of the long-standing observations about Archean cratons was their wide range of ages. Some cratons, like the Pilbara of western Australia, are as ancient as 2.8 billion years old; whereas others, like the North China craton on which I am sitting now as I write, are as young as 2.5 billion years old. Some cratons are even younger. Granted there was as much as 200 million years between the assembly of megacontinent Gondwana and the final amalgamation of the larger supercontinent Pangea. But this is nothing compared to the protracted time—almost as much as a billion years—it took for all the cratons to form. Even being very conservative in our definition of what a craton is, their formation ages span around 500 million years or more. So, just considering the timing between cratons and even before thinking about their spatial relationships, Bleeker realized that an Archean supercontinent may not even be a viable possibility. Ancient cratons like Pilbara or Kaapvaal in South Africa, for example, were intruded by a swarm of magmatic dikes. These dike swarms—the plumbing systems of volcanoes that are typically a geologic sign of continental breakup—were emplaced in Australia and South Africa around the same time that younger cratons like North

China were only starting to form. How could all or most of these cratons have been part of the same supercontinent if it had started to break up even before it was completely formed?

Bleeker had a different idea. He noticed that the earlier-forming cratons typically shared similar geologic histories and the later-forming cratons also shared similar geologic histories, even though these two or more groups contrasted so considerably if taken as a whole. Bleeker transformed these temporal groupings into spatial clusters. He called these clusters of cratons "supercratons."[26] He chose this term because, like continents that are rifted pieces of a supercontinent, cratons are rifted pieces of their supercraton. What if, Bleeker hypothesized, clusters of cratons had assembled together like younger supercontinents, but, unlike a supercontinent, there were multiple clusters, or supercratons, and they were geographically distant from each other and tectonically drifting separately? The multiple supercratons hypothesis could explain why some Archean cratons formed early and others late. It stood as a viable alternative to the Archean supercontinent hypothesis and, best of all, it could be tested.

•  •  •

I wasn't the only mentee that Bleeker brought with him up north. Bleeker's supercratons hypothesis was still global in scope even if it was competing with the possibility of a globe-girdling supercontinent. Just like the testing for supercontinents discussed in the past three chapters, testing Archean supercratons would require acquiring data from at least a few handfuls of cratons globally. The Geological Survey of Canada, based out of Ottawa, had already done a pretty knock-up job of studying the local geology of the Superior craton, named after Lake Superior, which defines the craton's northern margin. Restless Bleeker ventured, however, where few survey research scientists dared to go and found his field geology home, like

SUPERCONTINENT                    SUPERCRATONS

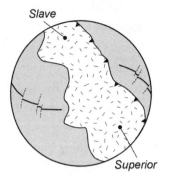
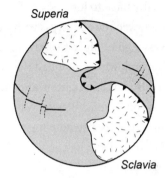

*vs.*

Figure 35. The competing hypotheses for Archean time. In the supercontinent hypothesis, all cratons are connected in a globe-girdling Archean supercontinent. According to the supercratons hypothesis, cratons are connected in clusters of supercratons, but these are segregated from each other and tectonically drifting independently of each other. Adapted from W. Bleeker, "The late Archean record: A puzzle in ca. 35 pieces," *Lithos* 71, no. 2–4 (2003): 99–134.

Hoffman and Bowring years earlier, in the tundra of the Northwest Territories to study the Slave craton, named after Great Slave Lake, the tenth largest lake in the world and the deepest in North America. (The egregious name for such a majestic place evidently comes from French fur traders, informed by the Cree tribe who were referring to the home of their enemy, the Dene tribe.) In addition to its alluring remoteness, Slave craton offered Bleeker a nice contrast with the Superior craton. There were substantial differences in the geologic histories of the two cratons, thus providing a natural setting to start testing his supercratons idea. If Bleeker was right, Superior had a cluster of cratons around it, which he called the putative "Superia" supercraton, and Slave was part of a different cluster of cratons that he called the "Sclavia" supercraton (derived from the Greek word for "slave," *sclavos*) (fig. 35).

Bleeker had also taken another young scientist under his wing and up to the Slave craton: Peng Peng from China. If Bleeker wished to test his own hypothesis with a global approach, he would eventually have to look beyond the confines of Canada, however well-endowed the country is with Archean cratons. Even though Canada was a good place to start, Bleeker had his eyes set on the horizon beyond, which included the Wyoming craton of the United States, the North China craton, and the Dharwar craton of India. By mentoring young Peng, Bleeker was making an ally in these grand plans. But Bleeker was also helping Peng. An up-and-coming young geologist in China, Peng had graduated from the esteemed Peking University and had in fact been a classmate of mantle convection modeler Nan Zhang. By the time Peng came to Canada to improve his English and learn Archean geology from Bleeker, one of the best, Peng was associate professor at the Institute of Geology and Geophysics, Chinese Academy of Sciences in Beijing. My meeting Peng in 2009 accounts for my currently very fortunate career as a researcher at this same institute in China. But back then, with both of us much younger, Peng and I had much to learn about doing field work far up north.

Having trained each of us independently, Bleeker decided to let Peng and me fend for ourselves together. We unfortunately got off to a rocky start (sorry) because the rock drill I had shipped up north was missing a critical piece, and one we could by no means acquire in the "city" of Yellowknife. We were at the ends of the Earth, truly where civilization stops and the wilderness begins. In fact, the "highway" we would take east of Yellowknife eventually just ends abruptly at a lake. This is where the "ice road truckers" would start their winter voyages to mining camps scattered about in the wilderness in the frozen season. Luckily, even at the ends of the Earth a geologist always has one trusty piece of equipment that never fails: a rock hammer (fig. 36). A rock hammer is similar to a normal hammer but has a little extra mass and sometimes a longer handle, giving it a little extra heft when trying to split a rock instead of merely tapping in a nail. Despite my best efforts at self-sabotage with my nonfunctional drill, Peng was fortunately a magician with a rock hammer, saving the trip and my PhD research.

Figure 36. Rock hammers all around in the Slave craton, Northwest Territories, Canada. (*Left*) From left to right, the author and Peng Peng. Photo credit: Wouter Bleeker. (*Right*) Wouter Bleeker inspects a thin dike with his rock hammer in his left hand. Photo credit: Ross Mitchell.

Minnesota may be known as the "land of 10,000 lakes," but northern Canada has thousands more. Studying geology in the Northwest Territories of Canada therefore involves crossing a lot of water between rock outcrops. Bleeker had trained me, with no prior experience, to be an expert in the inflatable Zodiac watercraft that we could store in the trunk of our field vehicle. Drive to the lake we needed to cross, inflate the Zodiac boat, mount the outboard motor, and speed on our way. Most of the sampling Peng and I had done up until that point was from the more easily accessible outcrops along the highway both east and west of Yellowknife. So it was finally time to start targeting the more remote outcrops out across the many and large lakes in the area.

We had spent a truly productive day of sampling in Wool Bay of Great Slave Lake. The target was the 2.6-billion-year-old Wool Bay diorite, a beautifully speckled rock similar to a granite but that had undergone less fractional crystallization. This was precisely the type of rock that could test Bleeker's supercratons hypothesis. First, the rock was the right Archean age. It could be, and had already been, precisely dated with U-Pb geochronology on its zircon crystals. Second, as a diorite, its intermediate composition somewhere between a granite and a basalt made it even better than a granite for paleomagnetism study. The Wool Bay diorite, rich in iron compared to a granite, had almost as much of the critical magnetic mineral magnetite as a basalt. The scene was set for a perfect day of sampling—and it had been perfect up until that point. But the one thing you can say about all geologic field work is that nothing ever goes to plan.

Peng and I had collected a bundle of diorite samples that day and were triumphantly on our way back home, boating back across the lake into the setting sun. Given the high northern latitude of Yellowknife (just shy of the Arctic Circle), the sun might have been setting at 9:00 p.m. or so at this time of the late summer. We of course took full advantage of this protracted daylight and made a full day (and night!) of it. Everything seemed to be going our way and according to plan. Then, the boat started to putter, and then, without much warning, it just stopped. The engine cut out and we came to a halt as our momentum died. We were close to the shore, but not as close as one wanted to be if they had to paddle a large Zodiac boat filled with rock samples with canoe paddles stored for emergency use only. We were in trouble.

As a student at the time, I did what students do: I went through the protocol, step by step. But the motor still wouldn't start. I didn't try it more than once or twice out of fear of flooding the engine. But each time, nothing. The protocol wasn't working. It was time for critical thinking—something that students are certainly taught, but certainly something that also comes from the experience of having to deal with problems that naturally arise over the course of one's life.

Peng had never driven the boat. He didn't even know the protocol. But he was able to do what I wasn't: stay calm. Even though he didn't have much (or any) experience with an outboard motor, as a good geologist, he was able to make observations. I had assumed it was an issue with the motor. So I had checked and double-checked the choke, the blade, and even under its hood. I couldn't find the problem.

As I became increasingly frustrated and upset with myself, Peng calmly pointed out that the gas can appeared to be crushed. It was sucked in, like it wasn't getting enough oxygen. I had missed a critical early step in the protocol. I had forgotten to recheck the airflow when I had gone through my checklist. The valve at the top of the gas can was indeed closed, instead of slightly open so that air could flow in. Without positive pressure from the air outside, the gas wasn't being pushed down the line to the motor. Peng opened the valve so the air could flow, and the once dormant motor started right up again and we were on our way. The beer and chicken wings (the closest thing that the late-night pizza place had to Chinese food) tasted especially delicious that night.

• • •

Those hard-earned diorite samples Peng and I collected would surely tell us exactly what we needed to know to test Bleeker's supercratons hypothesis. But to unlock the secrets of the samples we had collected, I had to make careful laboratory measurements on the magnetometer back at Yale.

To set the stage, the published paleomagnetic data from Superior and its clan of cratons suspected to be part of supercraton Superia yielded high-latitude results for this time (~2.6 billion years ago). Paleomagnetic poles for cratons of putative supercraton Superia all plotted essentially on top of those cratons. That is, using the paleomagnetic poles to reconstruct the ancient latitude, or paleolatitude, of the Superia cratons, one would rotate the poles to the North Pole and the cratons along with them. As the Superia poles each tended to plot on top of their cratons before the reconstruction was applied, when the poles were rotated to the North Pole, then their cratons came along with them, also ending up squarely in the frigid polar region.

The new data from Slave craton could not have been more different. You always remember when the first measurement rolls in. I held my breath. The first sample to be measured from the diorites that Peng and I had collected was nowhere near the pole. It wasn't even close to high latitudes. Goosebumps popped up on my arms and the back of my neck immediately. But then, almost as quickly, doubt crept in as I started to wonder whether the first measurement was reliable or just an anomaly. Will the other samples tell a similar story? Sure enough, one after another, all the samples began to show the same result: Slave craton was basking in warm tropical latitudes not far from the equator about 2.6 billion years ago. Most importantly, this surprising result implied that Slave craton wasn't anywhere near Superior and its clan of high-latitudes cratons.[27]

In fact, our data showed that Slave and Superior could not have been any farther away from each other. Given the great age of these rocks, we were unsure when we made our paleomagnetic measurements whether the magnetic direction measured was pointing to the North Pole or the South Pole. There is ambiguity in the polarity of the magnetic field, so we had to always consider the ramifications of both possibilities. If the poles measured for Slave and Superior were 180° away from other, this would indeed be the largest possible geographic separation two points on a sphere could have. However, because of the ambiguity in magnetic polarity, one could simply invert one of the two poles and their locations would actually have been identical. Therefore, the near 90° separation between the high-latitude poles of cratons related to Superior and the low-latitude pole of Slave (fig. 37) is in fact the farthest the cratons could have been from each other given the uncertainties of paleomagnetism in Archean time.

Such a large geographic gulf between the Slave and Superior cra-

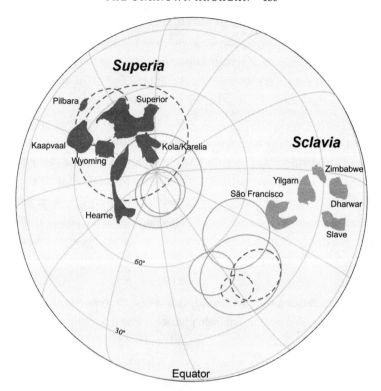

Figure 37. Archean supercratons as currently reconstructed. When the paleomagnetic poles are overlapped with each other at two different ages, the cratons end up in two clusters, or supercratons. The paleomagnetic poles of Superia are dashed and those of Sclavia are solid. Cratons, which are the rifted pieces of the few ancestral supercratons, are individually labeled. Adapted from Y. Liu, R. N. Mitchell, Z. X. Li, U. Kirscher, S. A. Pisarevsky, and C. Wang, "Archean geodynamics: Ephemeral supercontinents or long-lived supercratons," *Geology* 49, no. 7 (2021): 794–98.

tons was certainly not a ringing endorsement for the possibility of an Archean supercontinent. If a supercontinent in fact existed at this time, then all the other cratons we have yet to study would have to conveniently fill in the gap between these two clusters of cratons (fig. 37). Since my PhD, other cratons have been added to the picture, but they too seem to cluster around either Slave or Superior, and none have started to fill the gap.[28] It is conceivable that there were cratons filling the gap; but if these bits of crust ever existed, they must have been, for some reason, recycled back into the mantle and lost forever. Thus, although it remains difficult to rule out the possibility of an Archean supercontinent, the plausibility of its existence appears to be getting weaker instead of stronger as more data roll in.

Meanwhile, the isolated Slave and Superior cratons completely support Bleeker's supercratons hypothesis. Furthermore, the maps aren't just supported by paleomagnetism. The cratons that clustered around Superior all had dikes—sheets of magma feeding volcanic eruptions from deep in the crust—of precisely the same ages. Furthermore, these swarms of dikes had radiating patterns that all pointed back to the same places and that were always along the edge of the Superior craton.[29] These swarms formed because, as the ancestral landmass matured, much like a supercontinent, the mantle downwelling over which it formed had evolved into a mantle upwelling. Before long, mantle plumes rising from this deep mantle upwelling punctured the crust and breached the surface, creating huge outpourings of lava fed by dikes. When this ancestral landmass eventually broke up, it did so in the places where magma from these plumes had been focused.

But unlike a singular supercontinent, supercratons were multiple and segregated. The dike swarms of Slave and its cluster of cratons didn't match the dike ages of Superior and its cratons. The two clusters were distinct geographically and geologically. Despite this supporting evidence, the supercratons hypothesis is nonetheless still in its infancy. No hypothesis in science is accepted until all conceivable tests are conducted. But the scales seem to be tipping in its favor. If an Archean supercontinent hadn't formed, why not?

•    •    •

With Earth's mantle convecting at the grand scale it does today, forming a supercontinent is basically unavoidable. With the dominant

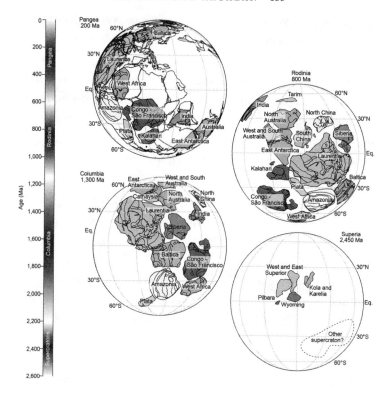

Figure 38. Supercontinents through time. Adapted from R. N. Mitchell, N. Zhang, J. Salminen, et al., "The supercontinent cycle," *Nature Reviews Earth & Environment* 2, no. 5 (2021): 358–74, published by Springer Nature.

degree 2 mantle structure—two opposite and equal hemispheres of upwelling, with only a thin ring of mantle downwelling between— continents are bound to disperse along the ring of fire, where eventually they will collide. Even today, with the continents as dispersed as they are along the ring of fire, we have the megacontinent of Eurasia already enlarged and showing no signs of stopping with Australia on a collision course to "soon" make it even larger, maybe as soon as 30 million years hence. And when Earth was characterized earlier by the even grander scale of degree 1 mantle structure, then sweeping all continents into the same place would have been even more unavoidable. At such scales of whole mantle convection, the formation of supercontinents is second nature. This is likely why we've had three supercontinents in the last 2 billion years (fig. 38). And with a fourth supercontinent on the way, the supercontinent cycle shows no signs of stopping. The quandary thus becomes what mantle convection looked like in the Archean to have potentially prevented an earlier supercontinent from forming.

Recall that computer models of mantle convection eventuate in the degree 1 mantle flow—one hemisphere of upwelling, one hemisphere of downwelling—even when they start from the humble beginnings of a random array of many small mantle upwellings and downwellings. Over time, one by one, downwellings and upwellings start to combine with their kind until only one of each is left. This evolving simulation is quite possibly what happened during the time between Archean supercratons and Earth's first supercontinent, Columbia.[30] If Archean mantle structure had not yet achieved degree 1 structure, then a supercontinent could not yet have formed. Furthermore, the multiple and geographically distant mantle downwellings could have been the foci over which Archean supercratons collected. But due to the presence of many mantle upwellings in between, these supercratons were effectively quarantined from each other, preventing their amalgamation into a large supercontinent. Earth's first supercontinent would thus have to wait for mantle convection. It had to wait for degree 1 flow to evolve, which took some time . . . potentially 2.5 billion years if Columbia, about 2 billion years old on a 4.543-billion-year-old Earth, is believed to be the planet's first supercontinent.

# 5

# THE NEXT SUPERCONTINENT

The difficulty lies not in the new ideas, but in escaping from
the old ones.

JOHN MAYNARD KEYNES

Before Damian Nance posited the Pangea conundrum, he was
working in Ohio with a whiz-kid colleague, Thomas Worsley, to
lead the charge in the modern era of supercontinent research.
They first talked about the occurrence of multiple supercontinents
through time, which they referred to as "Pangeas" plural, past and
future.[1] Then, again led by Worsley, they referred to a roughly
500-million-year-long "plate tectonic megacycle" and a "supercontinent assembly-fragmentation model."[2] Finally, they introduced
the term "supercontinent cycle" for the first time in 1986.[3] (Given
the manuscript was received by the journal in December of 1985,
the authors may have first uttered the phrase "supercontinent cycle"
eerily close to my ides of March birthdate earlier that year.)

A few years later, with Nance now in the lead spot, the supercontinent cycle was featured in the primetime pages of *Scientific American*.[4] The previous Worsley-led journal papers had been influential,
technically impressive, and advanced for their time. But the *Scientific
American* piece brought the nascent concept of the supercontinent
cycle into the limelight. This public-facing magazine is excellent

for attracting wide awareness of your work—which is also the most effective means to get other scientists to test your hypothesis. Years later, Paul Hoffman would take the same approach with the Snowball Earth hypothesis, following up his technical paper in *Science* with a more accessible article in *Scientific American*. Visibility and jealousy certainly fuel scientific hypothesis testing. Nance's team would continue to hone its model, but his group would no longer be alone in this pursuit. In fact, almost immediately others would propose an essentially opposite view of supercontinent formation.

• • •

In its first model for the supercontinent cycle, the Ohio team stipulated the following scenario: (1) Pangea started to break up, (2) the Atlantic Ocean opened up, (3) seafloor spreading made the Atlantic Ocean wider and older and, now projecting into the future, (4) subduction would initiate on either one or both sides of the old, dense Atlantic oceanic crust, (5) eventually the Atlantic Ocean would be consumed, and (6) the continents would collide to create the next supercontinent. And the cycle would repeat. In the Ohio model, the ocean that opens up during the breakup of a supercontinent is also the ocean that closes during assembly of the next supercontinent—so basically the Atlantic opens and closes over time.

According to this accordion-like model, a succeeding supercontinent would most likely somewhat resemble its predecessor. To be clear, the Indian Ocean also opened up during Pangea breakup. As we've mentioned before, the midocean ridges of the Indian and Atlantic Oceans form a continuous ridgeline. Although Nance's team might have emphasized "Atlantic-type" oceans in its publications, in the schematic figures of its accordion model the Indian Ocean was presumably included in this category of "interior" oceans that opened up during breakup (we too, for simplicity, may refer to the "Atlantic model" for such a scenario). Closing the interior oceans,

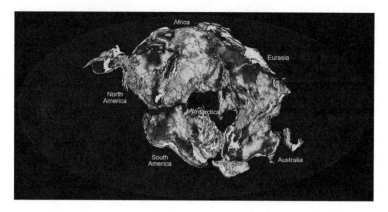

Figure 39. Pangea Proxima according to "introversion," or closure of the interior Atlantic Ocean, according to Christopher Scotese. Image credit: Pangea Proxima by C. R. Scotese, PALEOMAP Project.

with the exception of some continents possibly rotating a bit one way or the other, would more or less create a future supercontinent that would largely resemble Pangea and that would form in largely the same place.

Because the apple would not fall too far from the tree in their model, this might have been part of the reason the Ohio team referred early on to multiple Pangeas, past and future. Another advocate of such an Atlantic-oriented model is Christopher Scotese, who runs the Paleomap Project, which shows researchers the outcomes of their models. Scotese and other supporters of closing the interior oceans have offered numerous names for such a future configuration such as Pangea Proxima, Neopangea, and Pangea II (fig. 39).[5] Scotese even suggested Pangea Ultima to imply that he thought the next supercontinent would be Earth's last. But the argument for this is unclear.

Leaving a trail of peer-reviewed scientific literature to follow, the Ohio scientists provided logic for their Atlantic argument. They followed the Wilson cycle—the opening and closing life cycle of an ocean basin—discussed earlier (fig. 10). Recall that ocean basins are ephemeral and that the stages of their lifetimes are broadly predictable. In the Wilson-cycle-like scenario, the first three steps are simply history—what we know has already happened since the breakup of Pangea. Researchers start speculating only when it comes to the shrinking of the Atlantic—the subduction step. Subduction initiation remains a highly controversial topic, with many potential mechanisms at play.[6] In the late 1980s, Nance and colleagues thought the key to subduction was the old age of the oceanic crust at its continental margins. As we've discussed, as older oceanic crust gets farther away from the hot upwelling mantle at the midocean ridge, it becomes denser both because it cools and also because it gets increasingly hydrated over time by seawater creeping into its basaltic pore spaces. As we've also discussed, denser oceanic crust is an ideal candidate for subduction given its lack of buoyancy. In this sense, Nance's speculation—despite little actual evidence at the time for subduction in the Atlantic—didn't seem unreasonable.

Today advocates for this model point to two locations where small subduction zones exist, or even argue further that a third one is just beginning to initialize. Indeed, there are two bona fide, however small, subduction zones in the Atlantic Ocean, near the Lesser Antilles (known as the Lesser Antilles arc) and the South Georgia and South Sandwich Islands (known as the Scotia arc, where its namesake Scotia Sea was in turn named after a Norwegian ship that explored islands near the Antarctic Peninsula). Recall that magmatic arcs overlie places where subducting oceanic crust dehydrates and creates melting magmas that rise and form chains of volcanic islands like these. Most subduction on Earth occurs along the ring of fire encircling the Pacific Ocean. These two locations are in fact where Pacific subduction has encroached upon the otherwise tectonically quiet western Atlantic Ocean (fig. 40). But both these encroachments of Pacific subduction are restricted to latitudinal gaps between continents: the Caribbean subduction zone has encroached between North and South America and the Scotia subduction zone has encroached between South America and Antarctica. To close the Atlantic would require, however, subduction of large tracts of the Atlantic Ocean in longitude, hence bringing the continents back more or less to Africa from whence they came. This would require subduction to occur not merely in the gaps between the continents, but along at least one of the continental margins of the Atlantic Ocean. There is presently no evidence of subduction propagating along the continental margins of the Atlantic Ocean (fig. 40).

Atlantic advocates would nonetheless argue that there is at least suggestive evidence. Beneath the famously shallow Straits of Gibraltar between Spain and Morocco is yet another potential subduction zone in the realm of the Atlantic Ocean. Unlike the Lesser Antilles and the Scotia arcs, the Gibraltar arc is on the eastern side of the Atlantic Ocean. And there are faults forming in the Atlantic Ocean itself just west of the Gibraltar arc. Like a subduction zone where plates converge and compress each other, these eastern Atlantic faults are "thrust" faults where compression pushes one block up over the other. These thrust faults east of the Gibraltar arc are seen by some researchers as foreshadowing subduction initiating in the eastern Atlantic Ocean.[7] To be clear, there are indeed eastern Atlantic thrust faults, but is there is not yet a subduction zone. For the sake of argument, however, let's assume the Gibraltar arc someday indeed jumps west to these thrust faults and subduction initiates in the eastern Atlantic Ocean.

According to this scenario, theory advocates argue that there are three places where subduction is "invading" the Atlantic Ocean.[8] While "subduction invasion" is an interesting (and slant-rhyming) metaphor, we must consider the scientific mechanism that is implied if these three small raiding parties are to somehow conquer the whole Atlantic Ocean. Just like the Lesser Antilles and Scotia subduction zones occurring between their respective bounding continents, the Gibraltar arc also resides between two continents, Africa and Eurasia. Thus, as things presently stand, there is subduction occurring *between* the continents of the Atlantic Ocean, but not *along*

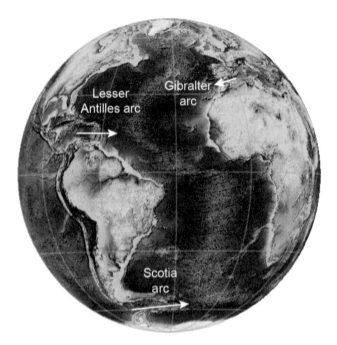

Figure 40. Introversion prediction for the closing of the Atlantic Ocean. The three arc systems are theorized to be the harbingers of subduction. Should such subduction initiate widely along Atlantic passive margins, it would someday consume the Atlantic. Presently there is no evidence for such subduction. Adapted from J. C. Duarte, W. P. Schellart, and F. M. Rosas, "The future of Earth's oceans: Consequences of subduction initiation in the Atlantic and implications for supercontinent formation," *Geological Magazine* 155, no. 1 (2018): 45–58.

their continental margins. This distinction might seem small, but it is potentially very insightful.

The subduction invasion model requires that these subduction zones somehow eventually propagate along the actual continental margins of the Atlantic. At present, however, there is no evidence that any of these three raiding parties have such grand goals. That is, none of these three subduction zones have branched out beyond their small tracts of oceanic crust sitting between the major continents. None of them has yet to show any signs of propagating *along* the continental margins—the real lines of defense that would have to be overcome to close the Atlantic Ocean. In the absence of evidence to support such a hypothesis, we can nonetheless assess its plausibility on theoretical grounds.

What are the forces that these invading subduction zones must overcome to subduct the Atlantic Ocean, and are their chances of success reasonable? Recall the main reason Nance's team envisioned subduction initiation at the continental margins of the Atlantic was because this was the oldest, and therefore the densest, seafloor in the ocean. While this may be true, there are other things to consider. It takes hundreds of millions of years for a large ocean like the Atlantic to open, with the Atlantic coming up on its 200-million-year birthday not long from now. The thickness of oceanic crust that is created during seafloor spreading at a midocean ridge is directly proportional to the temperature of the mantle. If the mantle is hotter, then more melt is generated and the resulting ocean crust is thick; if the mantle is cooler, then less melt is generated and the resulting oceanic crust is thin.

And indeed, this is exactly what the data show, and in every ocean basin, not just the Atlantic.[9] In the Atlantic, Indian, and Pacific Oceans, the oldest crust is the thickest and all younger ages get progressively thinner. But actually the thickness effect is in fact much more pronounced in the interior oceans (Atlantic and Indian) than it is in the exterior Pacific Ocean. This slight, but significant, difference is chalked up to the fact that the superheated mantle beneath supercontinent Pangea added some extra mantle heat that had to be dissipated, which wasn't present in the exterior realm of the Pacific. Thus, the most ancient crust of the interior oceans (Atlantic and Indian) is notably the thickest ocean crust on Earth. This interesting extreme turns out to be very unfavorable for initiating subduction on these most ancient continental margins of interior oceans.

Such continental margins are referred to as "passive margins" because there is no tectonic activity. Despite the boundaries between continental and oceanic crust, they are not plate margins. It is precisely at these passive margins that the accordion model of supercontinent formation would require subduction to suddenly appear. Is it feasible that such passive margins would suddenly become active? Probably not. Thicker crust is stronger. Stronger crust is less likely to bend or break. But bending and breaking are required for subduction to initiate. The old, thick crust at passive margins is also the strongest.[10] Even though it may be liable to subduct due to its density, as Nance and colleagues had originally envisioned, the passive margin needs to form a subduction zone in the first place. But there is even yet another reason why passive margins are very nonideal places to initiate subduction.

Passive margins, by nature, are both old and flanked by a continent to one side. Because of these two features, passive margins have extremely thick piles of sediment sitting on top of them. Sediment is shed from the nearby continent and deposited along the continental margin. The farther from the continental margin, the less sediment is deposited. Thus, that oldest ocean crust that formed when the ocean first opened, and that has always remained closest to the sediment-shedding continent, is covered with a thick shelf of sediment. The sediment loading of this continental shelf is heavy enough to put the underlying oceanic crust under stress—and not the kind of stress that helps initiate subduction. Loading the old oceanic crust with sediment imposes extensional stress, not the compressional stress that is needed for subduction to occur.[11] The stress that passive margins are under causes them to extend and is thus opposite to the kind that would help facilitate subduction.

Closing the Atlantic is a possibility that is difficult to reject, but the cards do seem stacked against such a bet. The notion that its oldest oceanic crust suddenly becomes dense enough to start spontaneously subducting when it reaches a certain "old" age is not supported by theory, nor is there any evidence that this is happening anywhere along either the eastern or western Atlantic continental margins. And interpreting the three (or two) subduction zones occurring within the oceanic crust between the continents rimming the Atlantic as an "invasion" that is a harbinger of more subduction to come is noteworthy, but only suggestive.

Before "invasion," the metaphor was "infection": "the introduction of subduction zones into a previously pristine ocean basin might be viewed, figuratively, as a process in which an 'infected' ocean basin comes into contact with, and infects, an uninfected basin." But good metaphors don't always produce long-lasting models. These same authors, when forced to come to terms with the Atlantic's stoic passive margins, also had to conclude that subduction cannot initiate at such passive continental margins simply because old ocean crust gets heavy.[12] As we will see, there other ways to interpret the intentions of these encroaching subduction zones that don't involve an overwhelming invasion of the entire Atlantic.

Quite naturally in science, when one hypothesis is proposed, there are other, alternative hypotheses that follow not far behind. This was certainly the case after Nance and colleagues proposed closing the Atlantic Ocean in a Wilson-cycle-like fashion. The Atlantic could indeed close—no one has a crystal ball—but we should also consider the alternatives.

•   •   •

It may at this point come as no surprise to you that Paul Hoffman would make one final appearance in our story—more surprising is that his cardboard cutouts do too. Sure, Hoffman's 1991 *Science* paper may have been crude in its execution compared to the GIS-based manipulation of geological data sets by Zheng-Xiang Li and his team years later. Nonetheless, in terms of Hoffman's conceptual thinking, what he achieved with his cardboard was nothing less than the next conceptual model of supercontinent formation. But Hoffman wouldn't do it alone.

Recall Hoffman's fan-like collapse of continents to create mega-continent Gondwana from the continents that had flanked centrally positioned Laurentia in supercontinent Rodinia (fig. 17). The placement of continents around Laurentia in Rodinia has certainly been adjusted since then (and continues to be so, particularly on Laurentia's enigmatic western margin).[13] Nonetheless, the basic precept remains: those continents that flanked Laurentia on either side during Rodinia would someday later unite in Gondwana. Hoffman being Hoffman, he skipped ahead to try to best describe the kinematics of the transition from Rodinia breakup to Gondwana assembly. But he had quietly challenged the prevailing notion of the Wilson cycle—that is, do oceans immediately close after they open up, or do some oceans wait around for some time before cycling on? South African geologist Chris Hartnady was there to notice the full ramifications of Hoffman's charging ahead.

Although common knowledge by the late 1980s, the Wilson cycle had unresolved issues, namely that fourth problematic step of subduction initiation—the troubles of which we have already discussed. These troubles indeed bothered Hartnady. In addition to the calculations of stress required to convert a passive margin into a subduction zone that called into question the feasibility of the Wilson cycle, Hartnady was also privy to the early attempts to reconstruct Pangea's predecessor in Precambrian time. The efforts of Hoffman, Dalziel, and others told in chapter 2 paved the way for Li's team effort and development of the modern conceptions of Rodinia. Everything Hartnady saw in these earliest reconstructions seemed to be at odds with the accordion-style Wilson cycle.

Hoffman's turning Gondwanan continents "inside out" from their positions in Rodinia to Gondwana sowed the seeds of Hartnady's doubt of the simple accordion model. Those continents that would become eastern Gondwana came from the eastern margin of Laurentia; and those continents that would become western Gondwana came from the western margin of Laurentia. But neither of these proto-Gondwana flanks of Laurentia in Rodinia broke away to then come back and collide once again with Laurentia. That was what Wilson's accordion model would have predicted, and it didn't happen in either case. On the contrary, those oceans that opened up when each proto-Gondwana flank broke away from Laurentia got wider and wider—never closing—and at the expense of the ocean that was exterior to them in Rodinia. That is, contrary to the accordion, Atlantic-type model that interior oceans open wide and then close back up, instead the interior oceans continue to widen and the exterior ocean closes.

Essentially what Hartnady had realized was that if one thought of plate tectonics as it occurs on a sphere, there was another possibility that wasn't apparent in the two-dimensional world of the Wilson cycle: death to the Pacific. He termed his alternative model supercontinental "extraversion," with the prefix chosen for its meaning "beyond" or "outside" the continents.[14] In addition to better explaining Hoffman's evidence of Gondwanan continents turning themselves "inside out," Hartnady also noticed an immediate theoretical appeal in his alternative model. Extraversion alleviated the weakest link in the Wilson cycle: subduction initiation in an Atlantic-type ocean basin. That is, no reversal of the Atlantic Ocean was needed if the Pacific could instead be closed.

Hartnady's model was still depicted in two dimensions, but in a two-dimensional cross section of the spherical Earth (fig. 41). This allowed one to see that the supercontinent could be formed without the continents collapsing back in on themselves, but instead by turning themselves inside out. Hartnady enjoyed the elegance of creating a new supercontinent without changing any of the margins from the breakup of its predecessor—unlike the Wilson cycle that somehow must convert breakup margins into subduction zones. That is, extraversion allowed the Atlantic Ocean that opened up with the breakup of Pangea to continue to spread, while the subduction zones originally encircling supercontinent Pangea continued to consume

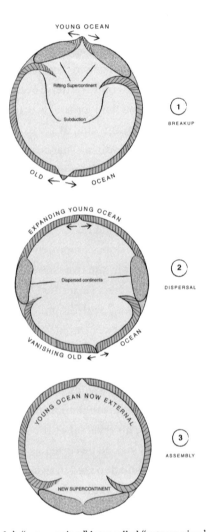

Figure 41. Hartnady's "extraversion," later called "extroversion," for "death to the Pacific." Unlike the Wilson cycle in 2-D, Hartnady took a 2-D cross section of the Earth that allowed for a distinction between interior and exterior oceans that alternate between supercontinent cycles. Note how the young interior ocean becomes an external ocean during the formation of the supercontinent by extroversion (closing external ocean). Adapted from C. J. H. Hartnady, "*On Supercontinents and Geotectonic Megacycles*," Precambrian Research Unit, Department of Geology, University of Cape Town, 1991.

the ocean crust of the exterior, Pacific Ocean. Extraversion could explain why the modern Pacific Ocean is shrunken in size compared to the larger Panthalassa Ocean that once encircled Pangea. But just because the Pacific might have shrunk over that past few hundred million years, does that prove it will entirely vanish?

Since Hartnady's pioneering work, a new generation of extraversion advocates have come along carrying on the battle cry "death to the Pacific." These additional considerations are certainly worth exploring, whether I personally subscribe to them or not.

• • •

Nance and Murphy of the Pangea conundrum also have a final appearance to make in our story. This time, they would take stock of the now two hypotheses of supercontinent formation we have discussed. They also made another critical, however simple, contribution: they gave the two competing hypotheses catchy names. New terms are no substitute for new science. But in this case, as in the case of Bleeker's coining of "supercraton," the new terms were a welcome addition as they simply gave names to the end-member hypotheses of supercontinent formation that people were already discussing. Murphy and Nance referred to the closure of the Atlantic-type interior oceans as "introversion" and the closure of the Pacific external ocean as "extroversion."[15] This was only a slight modification from Hartnady's extraversion, and it gave the competing hypotheses symmetry as well as humanizing the concepts by borrowing from the well-known personality types.

Death to the Pacific, or extroversion, is chiefly appealing because nearly the entire Pacific Ocean is rimmed with subduction zones—the ring of fire. At face value, then, what is there to stop the Pacific from vanishing in due course, since it is being consumed on all sides?[16] I argue this interpretation is a bit simplistic for several reasons. For starters, the East Pacific Rise (the midocean ridge akin to the Mid-Atlantic Ridge, but in the Pacific) represents the fastest seafloor spreading on Earth. Thus, even as subduction along the ring of fire is consuming the Pacific seafloor, the productive East Pacific Rise is replenishing the seafloor supply.

But then the extroverts would contend that the East Pacific Rise's days are numbered, pointing out that it someday "soon" (10–20 million years from now?) might be subducted. Indeed, the northern extension of the East Pacific Rise has already been subducted, the interactions of which gave rise to the modern-day San Andreas Fault and the jumbled geography of small transported tectonic blocks along the California coastline.[17] In addition, South America continues to move westward as it has for the past 120 million years.[18] It is this eastward subduction of the Pacific seafloor under westward-moving South America that has given rise to the looming magmatic arc of the Andes Mountains. Extroverts thus argue that it is inevitable that the East Pacific Rise will someday be subducted beneath the Andes too. It would indeed be dire to lose the East Pacific Rise, but would it really seal the fate of the Pacific?

Extroverts also like to point out the peculiar nature of subduction in the western Pacific to help make their case for a vanishing exterior ocean. Subduction zones in the West Pacific are notably different from the ocean-continent subduction zone, or continental arc, of the Andes. In the West Pacific, subduction occurs within the ocean—ocean is consuming ocean and the continents are far from the action. Such ocean-ocean subduction zones produce island chains like the Philippines and are called oceanic arcs (as opposed to continental arcs like the Andes). Even though these western Pacific oceanic arcs are actively subducting Pacific seafloor to the west, the location of the trench of the subduction zone is "rolling back" to the east. How is this possible? There are small oceans, or seas really, that open up to accommodate the rollback of the subduction trench.

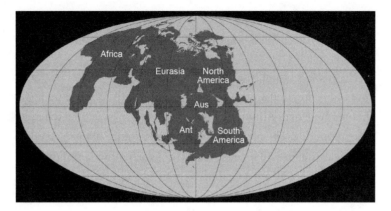

Figure 42. Novopangea formed by extroversion forecast by Roy Livermore. Adapted from Wikipedia with courtesy of the Creative Commons license CC BY-SA 4.0: https:// commons.wikimedia.org/wiki/File:Image_200.00my_(Novopangea).jpg.

Think the Japan Sea or the Tasman Sea. These narrow seaways are called "back arc basins" because they are behind the magmatic arc of voluminous granites produced above the subduction zone. So, going from east to west, one first encounters the subduction trench deep offshore, the island of Japan (the magmatic arc), the Japan Sea (the back arc basin), and finally mainland China. The Japan Sea continues to get wider as the trench rolls back to the east, even as the Pacific Ocean is subducted underneath Japan to the west.

This is thus the other reason extroverts see the Pacific as in the middle of a vanishing act. Not only is the Pacific being consumed in the east as the Atlantic Ocean continues to widen, pushing the continental arcs of North and South America farther west, it is also being consumed in the west as the oceanic arcs of the West Pacific roll back. Eventually, it is thought, the ring of fire will collapse in on itself.[19] Roy Livermore of Cambridge University has dubbed a supercontinent by closure of the external, Pacific Ocean, "Novopangea" (fig. 42). Such extroverts argue the Pacific is caving in, becoming vanishingly smaller on both sides for different reasons, but for effectively the same outcome: death to the Pacific. There are, however, several potential problems that extroverts must explain, as well as a possible way for the Pacific to survive.

No one can argue the fact that the Atlantic Ocean is actively spreading, and that spreading rates in the South Atlantic are faster than in the North Atlantic, allowing South America to catch up to North America. The present longitudinal distance between the two Americas that are offset east-west from each other is so large that the east coast of North America (New York) shares a time zone with the west coast of South America (Lima, Peru). In terms of longitude, because the South Atlantic opened up 40 million years after the North Atlantic, South America had some catching up to do. But that by no means portends an indefinite westward advance of the Americas to consume the Pacific Ocean. What if South America is just trying to catch up to North America, but will stop its longitudinal advance once it gets there? As we will see later in this chapter, it may even be possible that South America has already stabilized where it is, and that the Americas are arguably transitioning from their past days of westward advance to a future of simply rotating toward each other. In such a scenario, it's even conceivable that the East Pacific Rise may not necessarily die. But for argument's sake, let's just suppose it does.

I'll concede that consuming the East Pacific Rise would be bad for the Pacific, but it's not necessarily a death wish. Recall the back arc basins of the West Pacific. In many ways, it's hard to argue the difference between a back arc basin and a midocean ridge, other than their different widths. If the numerous back arc basins of the West Pacific were able to combine forces and link up together, the ridgeline of seafloor spreading running up and down the West Pacific would strongly resemble a newly formed midocean ridge. That is, even though the East Pacific Rise may inevitably disappear, it may very well have its successor forming and ready to replenish the seafloor of the Pacific Ocean if that fateful day comes. Such a nascent West

Pacific Rise could very well be the savior of the otherwise doomed Pacific Ocean.

These are the arguments both for and against the introversion and extroversion models. As a first pass, I have done my level best to assess the classical models on the playing field on which they were proposed and have been advocated for: as plate tectonic models about the fate of ocean basins.

• • •

Since their conception in the late 1980s (introversion) and the early 1990s (extroversion), advocates of each model, as we've discussed, have each invoked tectonic mechanisms for *how* each model would be achieved tectonically. But the question that lingers for both previous purely kinematic models is *why*. They tell us what goes where, but not why. Even with the tectonic mechanisms tacked on in later years—tectonic inversion and subduction invasion of passive margins for introversion and combined slab advance and rollback for extroversion—these models are still purely kinematic models. Their implied reasons for *why* are because either an interior or an exterior ocean should inherently close. Some researchers have even argued that *both* introversion and extroversion are possible and that maybe even the supercontinents have alternated from one mechanism to the other with each cycle.[20] If both models—literally opposites of each other—are to be true, then the enigma deepens.

Introversion and extroversion are not geodynamic models. That is, even with their tectonic descriptions, they still provide no link to mantle convection, which is a major factor in the dynamics of the supercontinent cycle. By this point in the book, you know that today's placement of continents was anything but random. Recall that Africa was the keystone of Pangea—the most central piece away from which the continents have dispersed.

At present, *all* of the continents (with the exception of Africa) are

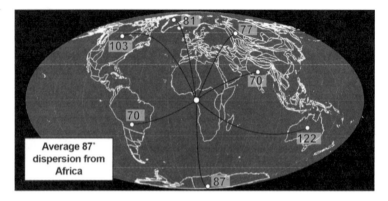

Figure 43. Orthogonal distribution of the present-day continents. The distances are measured from Earth's central axis under Africa (the minimum moment of inertia) to the calculated "centroid" of each continent. All continents, on average, are 87° away from Africa—the center of the last supercontinent Pangea and from which the present continents have dispersed. This systematic dispersion means that continents have gotten as far away from Africa as possible while also not getting too far into the Pacific Ocean: the continents are dispersed such that they are halfway between the African and Pacific plates.

positioned, on average, 87° degrees away from Africa (fig. 43). This average dispersion is strikingly close to 90°, that is, an idealized half hemisphere away from Africa. Another way of describing this angular relationship with which we are already familiar is that the continents are scattered along the Pacific ring of fire. That is, the continents are systematically positioned squarely between the African and Pacific plates. We tend to think of longitude in terms of eastern and western hemispheres with an arbitrary dividing line in Greenwich, England. But Earth actually has a natural longitudinal division between hemispheres: the African and the Pacific hemispheres, with all of the continents other than Africa being caught in between.

This systematic arrangement of the present-day continents is more obvious if we look at Earth in an unusual, but in this case, helpful projection: a transverse Mercator projection of Earth centered on the girdle of mantle downwelling and subduction between the

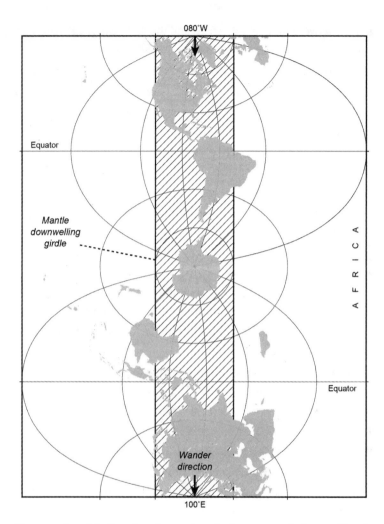

Figure 44. A straight line of continents along the degree 2 mantle downwelling girdle. Africa is not shown in order to emphasize that all other continents are distributed
orthogonal to (about 90° away from) Africa, the center of the last supercontinent Pangea. Orthoversion predicts that the continents are lined up because they are due to
collide with each other along this band. This is a transverse Mercator projection centered on the longitude of the direction of wander. This is done so that the edges are
where the two opposite axes of rotation for wander occur beneath Africa and beneath
the Pacific plate.

mantle upwellings beneath Africa and the Pacific (fig. 44). This perspective provides an easy way to see that all the continents (except Africa) run in a straight line bisecting Africa and the Pacific plate. That is, the continents have drifted as far as possible away from Africa without getting too close to the Pacific. They have found a middle ground: far away from Africa from whence they came, but not so far as to show any signs of overtaking the Pacific.

Why are the continents so systematically distributed? The 90° dispersion of the continents away from Africa can be described as each of the continents being "orthogonal," that is, positioned at right angles from Africa. This geometry inspired the name for my own preferred model of the next supercontinent: in contrast with introversion and extroversion, I have coined "orthoversion."

Each competing model of the supercontinent cycle predicts different ocean basins to close. Introversion would have the center of the next supercontinent be similar to Pangea, thus predicting a collapse back to Africa. Extroversion stipulates that the next supercontinent would be positioned ideally 180° away from its predecessor, predicting the continents to meet in the middle of the Pacific. Orthoversion is smack-dab between these two previous models, predicting that the next supercontinent will be positioned about 90° away from Africa.[21] As this band is where all the continents are presently positioned after their breakaway from Africa (fig. 45), orthoversion appears to be strongly supported by the orthogonal distribution of the present-day continents. But, whether you believe me or not, it was not this most obvious evidence hiding in plain sight on every world map I'd seen since grade school that inspired the orthoversion hypothesis. Science works in mysterious ways. Instead of rewriting history to make me aware of the obvious, I will rather take you along on the adventure that actually brought me to develop the hypothesis.

•   •   •

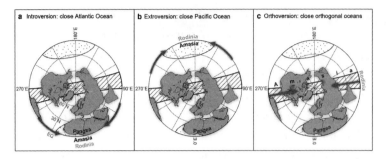

Figure 45. Supercontinent cycle hypotheses. Predicted locations of the future super-continent Amasia, according to three possible models of the supercontinent cycle: (a) introversion, (b) extroversion, and (c) orthoversion. The labeled centers of Pangea and Rodinia are the conjectured locations of each supercontinent's center. Equatorial cir-cles (stippled) represent supercontinent-induced mantle upwellings, and the orthogo-nal great circle swath (hatched) represents Pangea's downwelling/subduction girdle. In (c), Amasia could be centered anywhere along Pangea's subduction girdle. Arrows indi-cate where ocean basins would close according to each model. Continents are shown in present-day coordinates. Adapted from R. N. Mitchell, T. M. Kilian, and D. A. D. Evans, "Supercontinent cycles and the calculation of absolute palaeolongitude in deep time," *Nature* 482, no. 7384 (2012): 208–11.

I might have coined the term "orthoversion" and been lead author of the paper that fully proposed and tested the hypothesis. But it wasn't originally my idea. The idea was basically spelled out in a handful of prophetic sentences in a paper by my PhD advisor, David Evans of Yale University:

"The disaggregated fragments of the old supercontinent may drift into the subduction-downwelling girdle but advance no further; instead, they are trapped within the downwelling zones. . . . The next supercontinent assembles by collisions of these continental blocks within the downwelling girdle. . . . Every ca. 500 [million years] involves a ca. 90° shift in the fun-damental convective plan of mantle upwellings and down-wellings. Can we search the geological record for the sites of long-vanished supercontinent-induced upwellings?"[22]

Indeed, I thought we could. Reading of an inspiring new research direction in a paper by your PhD advisor is like receiving marching orders—and I followed them.

Evans painted yet another picture of the supercontinent cycle, but his vision went beyond the previous versions of Nance and of Hartnady. Recall that Nance and colleagues, following the Wilson cycle, had depicted the supercontinent cycle in two dimensions. Then Hartnady, realizing the importance of thinking of Earth as the sphere it is, depicted the supercontinent cycle in two dimensions again, but as a cross section of the planet that allowed a distinction between interior (Atlantic-type) and exterior (Pacific-type) oceans (fig. 41). Evans would be the first to depict Earth as it truly is, in three dimensions, and this made all the difference. One could try to get the basic point across in a Hartnady-like two-dimensional cross section, but adding the third dimension is a powerful illustration of what I would someday come to call "orthoversion."

The main key to Evans's 3-D depiction of the supercontinent cycle (fig. 46) is the dominance of degree 2 convection—what can be thought of as Earth's mantle like a convection oven with not just one, but two, heating elements opposite to each other. Recall that this is the structure that dominates the lower mantle today as well as the structure that is so closely associated with best-known super-continent Pangea. With Evans's picture in mind, we thus return to his gnomic utterances, one by one: "The disaggregated fragments of the old supercontinent may drift into the subduction-downwelling girdle but advance no further; instead, they are trapped within the downwelling zones." We start with a supercontinent having formed and having developed its underlying mantle upwelling in accordance with Zhang and Zhong's modeling. As continents break off and flee away from the sub-supercontinent upwelling, they would only go so far as the ring of downwelling between the two upwellings, but then go, in Evans's words, "no further." With the rifted continents "trapped" between the two hot upwellings, "the next supercontinent

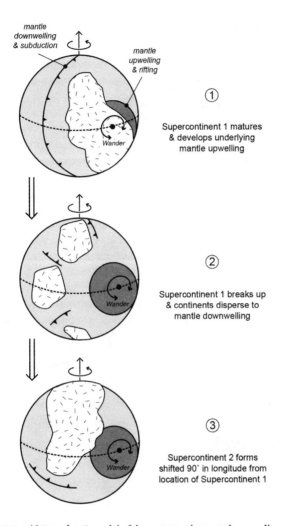

Figure 46. David Evans's 3-D model of the supercontinent cycle according to degree 2 mantle convection. With a 90° shift in longitude in the center of one supercontinent to the next, this laid the groundwork for developing the orthoversion model of the supercontinent cycle with Evans as my PhD advisor. The black lines with teeth represent subduction zones, the black dot represents the rotation axis of true polar wander, and the arrows at the top represent Earth's spin axis. Adapted from D. A. D. Evans, "True polar wander and supercontinents," *Tectonophysics* 362, no. 1–4 (2003): 303–20.

assembles by collisions of these continental blocks within the down-welling girdle." Evans was thus predicting that the next supercontinent would form 90° from the center of the last one. And the cycle continues—only shifted 90° in longitude.

Scientists must always make assumptions, and Evans was assuming the dominance of two-sided degree 2 mantle convection. According to his thinking, the next supercontinent would form 90° away from Africa—the center of Pangea still sitting atop the uplifted mantle upwelling that all of the continents of Pangea once sat atop. Degree 2 convection thus makes sense for Africa's central positioning and lack of plate tectonic motion over the past 200 million years. But does it make sense for all of the other continents? Evans's model is a nice concept, but how does it stack up to the Earth we have and know so well—down to the lower mantle—today?

The question of why the present-day continents are systematically positioned orthogonal to Africa can now be answered. The arrangement of the continents is controlled by underlying degree 2 mantle convection—the two-sided heating convection oven of the mantle under the African and Pacific plates has caused the continents to settle over the cold mantle in between. Once at the center of Pangea, Africa still sits on top of the hot mantle upwelling that developed beneath Pangea. Riding high on the upwelling, Africa is not pulled very much in any one particular direction or another. In fact, Africa is nearly completely surrounded by seafloor spreading ridges, and will be so even more completely if the East African Rift develops into a midocean ridge someday, as many expect it will. All the other continents have fled Africa by the opening up of the two interior oceans, the Atlantic and the Indian Oceans. Over the past 180 million years, these interior oceans have become progressively wider until the continents have ended up where they are today. Is the fact that all the continents are collectively dispersed 87° (nearly exactly 90°) away from Africa a coincidence? Or is it the expected outcome of some grand plan? According to orthoverts like myself and Evans, we maintain that it is the latter. That is, the 90° "orthog-onal" positioning of the continents around Africa is not only inter-nally consistent between all continents, it's a prediction of degree 2 mantle convection.

As the continents have drifted away from Africa, they have done so in a choreographed dance in accordance with underlying mantle flow. What again is so special about the 90° angle away from Africa? According to degree 2 mantle flow, with opposite and essentially equal upwellings underneath the Africa and Pacific plates, there is a bisecting girdle of downwelling in between, with the ring of fire where most subduction on Earth occurs. The continents are thus settling in this ring between the superheated and uplifted mantle upwellings. One might ask, nonetheless, how do we know that the continents' present-day locations nearly 90° away from Africa is not just a coincidence, a snapshot taken at this moment? How do we know that they will not just continue to disperse further and go beyond 90°? Such questions would likely be posed by extroverts—and in fact, I myself have encountered this question in peer review. Luckily, as the data would have it, the historical picture only further suggests that the present-day arrangement is not coincidental, but has arisen, continent by continent, over time.

Perhaps the best retort is found at the South Pole: Antarctica. Antarctica is more or less positioned squarely on the South Pole. In a world dominated by degree 2 mantle flow—two-sided mantle heating—this polar positioning is thus perfectly trapped between the both equatorial hot, uplifted African and Pacific upwellings. But as we've mentioned, this orthogonal positioning of Antarctica with respect to Africa is not particularly special—the other continents are too. What makes Antarctica special is not so much its position but its timing. Whereas other continents like South America are only just arriving at the degree 2 girdle, Antarctica got there nearly 100 mil-lion years ago. But, just as Evans predicted, it hasn't gone any farther since. It isn't invading the Pacific Ocean. It isn't returning to Africa.

It has stopped cold (sorry). The present-day orthogonal distribution of the continents is thus foreshadowing the future.

•    •    •

But making observations of the past few hundred million years only takes advantage of evidence from the Pangean and current super-continent cycle (the latter of which is still incomplete). However, Earth has had at least three complete supercontinent cycles in its past. Cannot these completed cycles provide multiple tests of how supercontinents form? Reading Evans's prophetic paper as a graduate student, I had an idea for how we could test his hypothesis of a 90° reset between supercontinents.

Our somewhat lengthy discussion of true polar wander, or simply "wander"—tumbling of the solid crust and mantle around the liquid outer core, toward the end of chapter 2—wasn't merely to justify my loss of limb. It was also to prepare us for the riches that come from keeping an open mind about polar wander. Wander is critical in understanding how supercontinents form as a result of whole mantle convection. As the motion is shared by all continents, it actually helps us figure out their tectonic configurations too. Wander is today almost as controversial as continental drift was during its day. It turns out that wander provides a critical way of testing the three competing models of the supercontinent cycle.

Wander rotates the whole solid Earth (mantle and crust) about an axis on the equator, but not just any axis. The physics of wander involve inertia, that physical laziness of mass to keep doing what it's already doing. Because we're talking about rotational motion in the case of wander, we use the term "moment of inertia," which is the tendency for a part of the rotating body to resist rotating faster or slower. By definition, wander is rotation occurring about an axis located where the moment of inertia is at a minimum. But where on Earth is this minimum moment of inertia and why?

Before thinking of Earth, in order to first understand what the minimum moment of inertia of a rotating body even is, think of a pencil. What is the easiest way to spin a pencil? With such an idealized object, you really only have two options and they couldn't be more different in terms of their moments of inertia. One way is easy, the other isn't. It's easy to spin a pencil about its long axis, that is, holding it on its tip and giving it a spin like a spinning top. Compare that simple, tight spiral of a rotation to trying to getting a pencil spinning end over end about its short axis. Typically, for most objects, the minimum moment of inertia is the long axis. So what's the long axis of the earth?

We come back again to the central importance of mantle convection and specifically its degree 2 structure (that is, two-sided mantle heating from both sides of the planet). The largest masses on the planet are the massive blobs sitting in the lower mantle, the two mirror-image LLSVPs (remember that egregious abbreviation from chap. 1? large low shearwave velocity provinces; fig. 13). These two masses and their associated convective mantle upwellings define the long axis of Earth. We measure wander very precisely today with satellites, and the equatorial axis about which wander occurs pierces squarely through Africa on one side of the earth and the Pacific Ocean on the other side of the planet. It is smaller masses associated with sinking slabs at subduction zones and rising mantle plumes that incite wander, requiring the planet to find a new stable axis around which to spin. But it is the minimum moment of inertia, controlled by the two opposite and equatorial LLSVPs, that is the axis about which wander reorients the planet.

But what is important to us here is that the wander axis thus pierces through the heart of the past supercontinent: Africa. Now we can return to Evans's final cryptic but inspiring clarion call of a question: "Can we search the geological record for the sites of long-vanished supercontinent-induced upwellings?"[23] Long-vanished supercontinent-induced upwellings, or ancient LLSVPs, would tell us where past supercontinents had been positioned. As the three competing models of the supercontinent cycle made such contrasting predictions for the longitude of successive supercontinents (fig. 45), finding ancient supercontinent centers and measuring the angles between them could test the three hypotheses.

It was thus critical to find a way to identify the centers of supercontinents. Clearly with the recent Pangea supercontinent cycle, we have several ways available to us to approximate the center of the now vanished supercontinent. For one, there is the African LLSVP sitting in the lower mantle. Reconstruct Pangea at the time just before its breakup and the supercontinent sits squarely on top of the African LLSVP. The LLSVP, or supercontinent-related mantle upwelling, is a great ancient indicator of where the now vanished supercontinent used to be centered. Unfortunately, however, this mantle structure is merely a snapshot of the present day provided to us by seismology. We will never have seismically inferred images of the mantle structure for more ancient supercontinents. So we must look elsewhere to locate the centers of pre-Pangean supercontinents.

This is where wander comes to the rescue. We have paleomagnetic data throughout much of Precambrian time. Otherwise we wouldn't have a quantitative way of making the reconstructions of supercontinents Rodinia and Columbia and Archean supercratons. But not only does paleomagnetic data measure relative motion of the continents due to plate tectonics, it also measures wholesale motion of all the continents due to wander. Lucky for us, the degree 2 mantle flow pattern that a supercontinent creates also makes Earth take a peculiar shape that is prone to become rotationally unstable and thus to wander.

Earth is already not a perfect sphere because its rotational flattening causes it to bulge around the equator. But then add the extra shape of the LLSVPs and their associated convective mantle upwellings, and Earth's figure bulges even more along these two portions of the equator, or along a particular equatorial axis because the LLSVPs

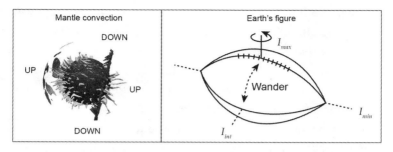

Figure 47. Wander and the shape of the earth. (*Left*) Two-sided heating, or degree 2, mantle convection. Earth's core is the spherical center, overlain by the two mantle upwellings (light gray) and the bisecting ring of mantle downwelling (dark gray). Adapted from R. N. Mitchell, N. Zhang, J. Salminen, et al., "The supercontinent cycle," *Nature Reviews Earth & Environment* 2, no. 5 (2021): 358–74. (*Right*) American-football-like shape of Earth due to degree 2 mantle flow. Wander occurs about the minimum moment of inertia ($I_{min}$), or long axis of the earth. Because the other two axes ($I_{max}$ and $I_{int}$) are nearly the same, any small addition or subtraction of mass will cause wander very easily when Earth is this shape due to supercontinent formation. Adapted from D. A. Evans, "True polar wander, a supercontinental legacy," *Earth and Planetary Science Letters* 157, no. 1–2 (1998): 1–8, and P. F. Hoffman, "The break-up of Rodinia, birth of Gondwana, true polar wander and the Snowball Earth," *Journal of African Earth Sciences* 28, no. 1 (1999): 17–33.

are opposite each other. Degree 2 mantle flow thus makes Earth's shape closely resemble an American football or a rugby ball, with the long axis setting the minimum moment of inertia around which wander rotates (fig. 47). Add or take away a little mass anywhere on the planet and, due to this football-like shape, the world will spiral around the long axis. Paleomagnetic data detecting this wander rotation would thus circumscribe a line of paleomagnetic poles. Simply fit a line to those data and the minimum moment of inertia is defined at a right angle to that line of data. This minimum moment can quite precisely approximate the center of each supercontinent. The test was all laid out: using the wander legacies of successive supercontinents Pangea and Rodinia, find their respective minimum moment axes and measure the distance between them.

I suppose it was this final intuitive leap that I made that Evans hadn't thought of: that we could use wander to test his idea of a 90° reset between supercontinents. We could use wander to test orthoversion—and all of the three models of the supercontinent cycle for that matter. For the data to support orthoversion, we would expect there to be a 90° shift from the wander axis of one supercontinent to that of the next one. As luck would have it, several critical paleomagnetic papers interpreting sudden and oscillatory rotations as wander had been published in the years just before my PhD. I was already intrigued by wander as a potentially underappreciated process on Earth, and these papers gave me the extra boost of confidence I needed to trust those instincts and see them through.

•   •   •

I remember the day I brought the first plots of the data to Evans's plush Yale office. He sat there quietly on his leather couch that had so many books and papers and geologic maps on it that it only left room for one, him, to sit. And so I stood there, holding up my laptop for Evans to inspect the outcome of the test of his own hypothesis. The result was almost exactly as he had predicted. The difference between the wander axis of Pangea and that of its predecessor, Rodinia, was 87° (fig. 48). Given the uncertainties involved, the result basically could not have been closer to 90° (orthoversion), and we could certainly reject both 0° (introversion) and 180° (extroversion). A supremely skeptical scientist, even when his own hypothesis is at stake—indeed, perhaps *particularly* when his own hypothesis is at stake—Evans admitted the data indeed appeared to fit his model. Whether he was quietly skeptical or just shocked, or some combination of the two, I couldn't tell. I reminded Evans that this was his idea after all. While flattered with the honest accreditation, Evans admitted to me then—having just seen the first successful test—that

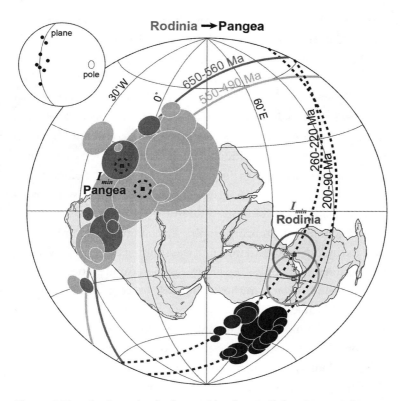

Figure 48. Test of orthoversion for the transition from Rodinia to Pangea. Paleomagnetic poles for each wander legacy of each supercontinent are lighter shades for Rodinia and darker shades for Pangea. Lines were fit to each wander legacy. (Lines on a sphere are called "great circles," i.e., the shortest distance between two points on a sphere, with the inset showing how a best-fit great circle, or plane, has a pole ~90° away.) The wander axes associated with each linear fit are labeled "$I_{min}$" for the minimum moment of inertia for each supercontinent. The two successive $I_{min}$ axes are nearly 90° apart from each other, as predicted by orthoversion. "Ma" means million years ago. Adapted from R. N. Mitchell, T. M. Kilian, and D. A. D. Evans, "Supercontinent cycles and the calculation of absolute palaeolongitude in deep time," *Nature* 482, no. 7384 (2012): 208–11.

he had just assumed there was no way to test the idea, however much theoretical sense the conceptual model had made to him at the time.

I had done my job of convincing Evans that his idea of a 90° reset between continents could be tested with data, and the kind of data with which we were experts, paleomagnetic data. Impressed with this newfound testing ability, Evans's skepticism kicked back in as he reminded me that there was another supercontinent transition that could provide an additional test. It was indeed too early to celebrate. I had more work to do. I had to test to see if the older transition between supercontinents Columbia and Rodinia also followed a similar pattern.

And so I was off, probably running, to conduct the additional test ordered by Evans. At least my mind was running. I was already trying to envision the outcome even before I gathered and plotted the data. From the dim vision in my head, the likely outcome seemed—to my imagination at least—likely to favor our 90° hypothesis. I was naturally excited, but I still had to make the cold calculations. A critical portion of the data we would be using for this older test came from Adam Maloof's hard-fought sample collections from the ice-covered limestone cliffs of Svalbard. This had been a critical paper in making the geology community really take wander seriously in recent years.[24] Long after he once incorrectly identified halite as pyrite, Maloof, forged in the crucible of Hoffman's supervision, had made fundamental contributions as a tenured Princeton professor. His Rodinia-aged data would be critical in our final test of orthoversion.

I nervously typed Maloof's data into the spreadsheet. After gathering Maloof's data for Rodinia wander, I turned to compiling the older Columbia wander data. Not surprisingly, Maloof had also played a hand in shoring up those data too. Many of the Columbia-aged paleomagnetic data came from Laurentia, specifically the "Mid-Continent Rift." You heard that right. About 1.1 billion years ago, North America almost split in half and the vestiges of this ancient rift can still be seen today in the shape and depths of Lake Superior of the Great Lakes. But the rift failed. Nonetheless, while it lasted, the Mid-Continent Rift, much like the East Africa Rift today, melted a lot of mantle which gave rise to a plethora of iron-rich basalts at the surface, ideal for acquiring paleomagnetic data.

Early paleomagnetic studies of the Mid-Continent Rift had noticed an odd phenomenon. Data of almost the same aged rocks yielded strikingly different paleomagnetic poles. The first papers on this, including one published in *Nature*, attributed the bizarrely large difference in directions from rocks of nearly the same age to a bizarrely behaving magnetic field.[25] Maloof, and his PhD student at the time, none other than my heroic first responder, Swanson-Hysell, found a clever way to prove that in fact the magnetic field was indeed behaving as usual—instead, it was the fact that Laurentia was moving surprisingly rapidly at that time that caused the paleomagnetic poles to shift.[26] By breaking the plate tectonic speed limit, the Mid-Continent Rift data, Evans and I suspected, had been influenced and expedited by wander. Thus, yet another Maloof data set proved crucial in our critical final test of orthoversion.

The spreadsheet was done and it was time to face the music and make the final plot. I plotted up both eras of wander, the wander legacy associated with Rodinia (the Svalbard data) and the wander legacy associated with Columbia (the Mid-Continent Rift data). The wander legacy of each supercontinent circumscribed a line of paleomagnetic poles, just as expected as wander oscillates the earth back and forth about a stable axis on the equator. As before, I held my breath as I selected the wander data of each supercontinent and fit lines to each of them. The lines were fit and each of their associated wander axes popped up. The two minimum moments of inertia axes appeared. By my eye, they looked just about right—about 90° away from each other, that is. But I still had to measure them to be totally sure. The angle between the wander axes of Columbia and Rodinia was 88° (fig. 49). Evans was now convinced and it was time to write the scientific paper. It would eventually be published

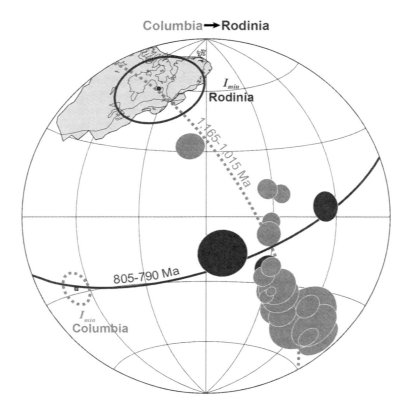

Figure 49. Test of orthoversion for the transition from Columbia to Rodinia. Paleomagnetic poles for each wander legacy of each supercontinent are lighter for Columbia and darker for Rodinia. Lines (or great circles) were fit to each wander legacy. The wander axes associated with each linear fit are labeled "$I_{min}$" for the minimum moment of inertia for each supercontinent. The great circles and $I_{min}$ axes for each wander legacy for each supercontinent are dashed and solid lines for Columbia and Rodinia, respectively. The two successive $I_{min}$ axes are nearly 90° apart from each other, as predicted by orthoversion. Adapted from R. N. Mitchell, T. M. Kilian, and D. A. D. Evans, "Supercontinent cycles and the calculation of absolute palaeolongitude in deep time," *Nature* 482, no. 7384 (2012): 208–11.

in February 2012 in *Nature* and represents the defining achievement of my career to date.[27]

•   •   •

Now that we have the evidence from the past—that orthoversion had happened between both of the two past supercontinent transitions—we are ready to think about what it would imply for the future. History repeating. How would a next supercontinent formed by orthoversion look different from Pangea Proxima (introversion) or Novopangea (extroversion)?

But before we launch into differences, let's start with the only common ground shared by all: Australia. The tectonic plate in which the Australian continent resides is a real outlier. As a general principle, purely oceanic plates without continents embedded in them (think the Pacific plate) move fast (more than 5 cm per year) and those plates with continents move slow (less than 5 cm per year). But the Indo-Australian plate breaks this rule. It is the only continental plate moving as fast as the oceanic plates. It is doing so because it has a great deal of northward subduction occurring along its northern margin. Because its current trajectory is already northward, this northward subduction adds an extra pull and so the continental plate moves quickly. But Australia is not only an outlier in terms of its speed.

Location, location, location. Recall that the present-day arrangement of the continents is such that they have systematically scattered to be orthogonal to Africa, thus settling down along the degree 2 mantle downwelling girdle. But not all continents had to travel the same distance to get there. South America, only just arriving now at the girdle, started snugly close to Africa and had to travel a vast 60° or so to reach the girdle. But India and Australia, effectively treated as a single plate despite their minor differences, had much less distance to travel to get to the girdle. When Pangea started to breakup 180 million years ago, India and Australia collectively were 71° away from Africa, that is, very close to the girdle already. Over time, they have settled even more on the girdle (presently 96° away from Africa), but they have basically nearly been there from the start.

The pump was thus already primed to fuel the rapid Indo-Australian tectonic motions, and that's exactly what both continents have done. Being positioned already on the girdle, the two continents have moved almost purely in latitude *along* the longitudinal girdle, both with rapid northward motions. There is thus little doubt, even between advocates of contrasting models of the supercontinent cycle, that Australia is currently on a collision course to hit Asia, just like India before it. The long-term trajectory likely has Australia joining the growing megacontinent of Eurasia somewhere between Japan and India about 30 to 40 million years from now.

But even though there is consensus on this most imminent collision, what does it tell us about the competing models of supercontinent formation? Which model is Australia's trajectory most consistent with? As far as extroversion goes, adding Australia to Eurasia is basically the least incursion on the Pacific that one could ask for and it is thus certainly not compelling evidence of the Pacific closing. On the contrary, in addition to moving north, Australia has moved a bit east, but only so far. As soon as Australia started to graze the Pacific mantle upwelling, it largely stopped moving east and has concentrated its efforts to the north. It thus shows no signs of wanting to climb up this mantle upwelling, nor should it. Similarly, Australia moving north has no bearing on closing the Atlantic. Australia's intentions are inconsistent with extroversion and irrelevant to introversion. Meanwhile, its northward motion along the degree 2 girdle, on which it was already positioned before other continents, is completely consistent with orthoversion.

•   •   •

But what about the fates of continents that are less clear? Hartnady, one of the original extroverts, made a relevant point about South America. He made an argument for why so-called subduction invasion in the Atlantic did not actually portend the closing of the entire ocean, but something else entirely:

> There is absolutely no justification, in terms of forward extrapolation of present-day plate-tectonic rates, for supposing that South America will be rejoined to Africa in a future Pangea some 200 million years hence. It may rather continue to rotate northwestward, pivoting around the Caribbean plate, until it collides with the Pacific, south-western margin of the North American plate.[28]

But Hartnady had not only taken a cogent stand against introversion. He had also fused the Americas, in much the same way that in fact orthoversion predicts they would be fused. Both models, extroversion and orthoversion, have South America rotating around to collide with North America. But orthoversion does two things slightly differently from Hartnady's description.

The main difference between Hartnady's vision of the Americas fusing together and that of orthoversion, is that orthoversion has a strict rule for longitude based on mantle convection. Hartnady would have the Atlantic Ocean to continue to spread indefinitely—or at least until it grew so wide that it closed the Pacific on the other side of the world. But according to this theory, it would have to be entirely a coincidence that the Americas, North and South taken together, are presently 89.9° away from Africa. Starting off 40° away from Africa when Pangea started to break up, they have taken the same path—the opening of the Atlantic—to get where they are today. If they planned to carry on their merry way across the Pacific, then their having reached 90°—the degree 2 mantle downwelling girdle—would make our observing this at the present-day a complete coincidence. But as with Antarctica at the South Pole, it is more likely a convergent solution, a recent achievement of the Americas that had been in the works for quite some time.

Because the Central Atlantic opened up first, North America has slightly overshot the girdle; and because the South Atlantic opened up second, South America may come up just shy of the girdle. But taken together as part of the Atlantic Ocean, the Americas have positioned themselves precisely over the girdle over the past 180 million years. And the reason the Americas have traveled so far for so long to reach the girdle—to settle in its topographic low—is the same reason we do not expect either of them to leave the valley now that they've arrived. Why flee down the mountainside of the African mantle upwelling only just to then ascend up the Pacific upwelling? Sure, anything is possible. But when trying to forecast the future, it's best to deal in the probable.

Whereas Hartnady had South America pivoting around the Caribbean Sea to collide someday with North America somewhere over the Pacific, orthoversion stipulates that the Americas will fuse by closing the Caribbean Sea. Located squarely on the girdle, the Caribbean Sea, according to orthoversion, is not long for this world. Maybe the continents are waiting for everyone to get to the girdle before moving along it. Recall that taken as a whole, all the continents dispersing away from Africa are presently 87° away—so close, yet so far. Extrapolating the trend of the past 180 million years since the breakup of Pangea, we may expect the fleeing continents to collectively reach the girdle about 30 million years in the future. This is arguably the first step of orthoversion—all of the continents reaching an orthogonal position—before they can proceed any further with assembling the next supercontinent. Once all continents reach the girdle, then motions along the girdle will likely ensue.

•   •   •

But how do we know continents intend to move along the girdle once they've reached it? Well, some have already started doing so. Remember the Indo-Australian plate that was already positioned quite close to the girdle much earlier on. India only very recently shot northward along the girdle, colliding with Eurasia about 50 million years ago, and Australia's present northward motion is set to have it collide with Eurasia at essentially the same time as the remainder of the continents reach the girdle, about 30 million years hence. Orthoversion therefore appears to be a two-step process: (1) get to the girdle and (2) move along the girdle.[29]

So who are the stragglers that the second step of orthoversion must wait for? The Atlantic and Indian Oceans have already done their job. As we've mentioned, taking the Americas as an Atlantic unit and the Indo-Australian plate as a unit of the Indian Ocean, both of these oceans have opened so that their respective continents are on the girdle. Those continents that fall short are Greenland and Eurasia. Will they finally get there? And if they are indeed destined to someday arrive at the girdle, what might push them to finally get there? As we are discussing the continued dispersal of continents away from Africa, the best clue might come from Africa itself.

The East African Rift is the only place where continental rifting is actively occurring on Earth today. This present-day novelty is surely telling us something about the immediate future. And this part of the Wilson cycle, steps 1 through 3 (fig. 10), is relatively uncontroversial. Most people accept that the East African Rift will continue to develop into the opening of an ocean. Indeed, the nearby narrow inlets of the Red Sea and the Gulf of Aden may already be just that, forming a "triple junction" where these two incipient seas meet the East African Rift. Assuming that Africa will continue doing what it's always done—sit comfortably on top of its underlying mantle upwelling—then seafloor spreading developing along the African arm of the triple junction will push the Arabian plate, and in turn Eurasia, giving it the final push it needs to get 90° away from Africa.

And what about the other straggler, Greenland? It may be closer to 90° than Eurasia, but it may take just as long to get there—it will be a race to the finish. Even though Greenland's already farther along, it may take just as long to reach the girdle as Eurasia because it's relying on, no joke, the slowest-spreading midocean ridge in the world: the Gakkel Ridge.[30] You may recall that this Soviet-predicted and Soviet-discovered ridge is the northern terminus of the Mid-Atlantic Ridge. The North Atlantic was the last portion of the Atlantic Ocean to open up, and it is only slowly, however steadily, catching up. It may take some time, but the Gakkel Ridge will eventually get Greenland to the girdle. Then phase 1 of orthoversion will be achieved globally and phase 2—continental collisions along the girdle—will be well underway and the wave of the future.

The other difference between orthoversion and Hartnady's extroversion is that the two models predict opposite coastlines of the Americas to collide. By pivoting around the Caribbean Sea, South America, Hartnady thought, would have its Andean margin and North America's Rocky Mountain margin collapse in on each other somewhere over the Pacific. Recall that even though the Andes and the Rockies have different names, they are still a part of the Cordillera, a continuous range of mountains running down the entire west coast of the Americas. Why these portions of effectively the same orogen that have worked in concert for so long would suddenly turn against each other someday in the future is unclear.

Orthoversion, on the other hand, would use the already active faults bounding both the north and south margins of the Caribbean plate to allow North and South America to rotate their east and west coasts, respectively, to face each other. New York City meet Lima, Peru. The long-term goal of the Lesser Antilles arc, according to orthoversion, would not be that it is destined to invade the Atlantic (introversion), but merely that it will stay where it is until the Americas are ready to migrate along the girdle on a collision course leading to their merger. As it is the coupling of the continents with underlying mantle flow that seems to make the big tectonic decisions in the grand scheme of things, and the oceans merely follow orders, I have my bet on the motions of the Americas along the girdle consuming the Caribbean Sea instead of pivoting around it (fig. 50).

Other differences from Hartnady's vision continue farther into the future, after the Americas are sutured. Hartnady would have the Atlantic continue to widen at speeds similar to modern rates. This would result in the fused Americas eventually colliding with Asia, which they would meet in the middle of the Pacific Ocean and thereby consume. Also an extrovert, at least originally, Hoffman actually coined the name "Amasia" for the meeting of the Americas and Asia according to extroversion in the Pacific hemisphere. Although orthoversion has adopted Hoffman's excellent branding skills, it predicts the Americas and Asia will indeed meet, but in an entirely different way and for entirely different reasons. Again, the main distinction is geodynamics, not merely geometry. (Co-opting Hoffman's name, Amasia, for the next supercontinent formed by orthoversion is hopefully justified by the fact that Hoffman is fond of the theory's foundation in geodynamics, not geometry.)

According to orthoversion, the Americas would collide with Asia by meeting at the North Pole. To do so, the Arctic Ocean would have to be closed (fig. 50). Is this reasonable? "Consistency should be admired," Mr. Perkins, my precalculus teacher, once told me. Consistency in a theory is a basic requirement for its making testable predictions. The reason the Arctic Ocean is predicted to close is the same reason the Caribbean Sea is doomed according to orthoversion: they are both situated along the girdle. Consistency. Once Greenland on one side of the Arctic Ocean and Eurasia on the other have both finally completed phase 1 of orthoversion—getting to the girdle—phase 2 will commence by starting to subduct, and someday close, the Arctic Ocean. To summarize, then, Amasia to this point will have formed according to the same principle of closing those seafloors situated along the girdle: Australia by following the way of

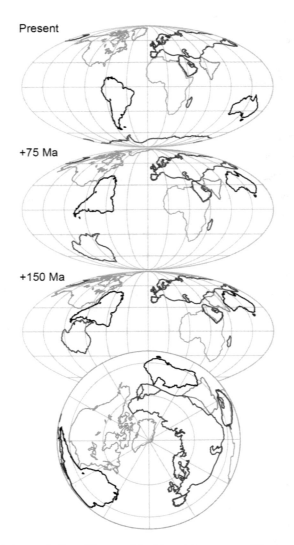

Figure 50. Future projection of the assembly of Amasia in 150–200 million years based on orthoversion. Bottom is Amasia as viewed from the North Pole. Rotation parameters for the continents were provided by David A. D. Evans.

India before it, both thus joining Eurasia; the closure of the Caribbean Sea to unify the Americas; and closure of the Arctic Ocean to merge the Americas with Eurasia. Amasia according to orthoversion is nearly complete.

•   •   •

Finally, there's the prickly question of Antarctica. At the time of our publication, my colleagues and I were agnostic about the fate of Antarctica. After all, it's on the degree 2 downwelling girdle. Whether it goes one way or the other doesn't matter for our orthoversion hypothesis. If it goes north via the west, the way of South America, that would be consistent with orthoversion. If were to go north via the east, the way of Australia, it would also be consistent with orthoversion. Either way, Antarctica had two reasonable paths to join Amasia. We had no reason to make a judgment call at the time. And one tactic I learned from David Evans (and am still trying to learn) is not to give picky peer reviewers too much to criticize. "One idea, one paper," I've been told by both Mark Brandon of the crossword puzzle analogy as well as by Brendan Murphy himself. So we tactfully left Antarctica out of the picture at the time.

But as a long-term strategy, the omission of Antarctica wasn't sufficient. According to a well-researched website (at least on this matter of which I am aware), Speculative Evolution Wiki (SpecWiki), this omission was orthoversion's Achilles' heel: "Compared to other supercontinents, Amasia is a contender for the most plausible. However, one problem with the Amasia layout is that Antarctica barely moves [and,] given enough time, will be forced to move north due to plate tectonics."[31]

To their credit, the writers at SpecWiki well understood our reason for Antarctic agnosticism. They explain, "a possible reason for the static layout of Antarctica is that not enough information is known to be sure which direction it will travel in the future. However, if Amasia occurs sooner (as in 50 to 100 million years from now), rather than later, Antarctica will likely not be a part of the supercontinent."[32] And in 2012, on radio interviews with *Nature* and *NPR*, that was just how I left it when asked about the southern enigma, saying that Antarctica may very well remain "stranded" at the South Pole. There was indeed some scientific literature available at the time that gave me reason to speculate so. But I'm not so sure anymore.

As of 2012, a seismology study had imaged a mantle plume in the upper mantle beneath Antarctica.[33] Such an upper mantle plume can explain why Antarctica is positioned so perfectly on the South Pole— because wander shifted the entire planet to bring it there. Just like the geyser shooting squarely out of the South Pole of Saturn's moon Enceladus, shallow plumes are more rotationally stable for a planet or moon if they are near the spin axis.[34] So both Earth and Enceladus have reoriented themselves by wander to bring these shallow plumes precisely to their poles. Subsequent seismological studies of Antarctica continue to confirm the presence of this shallow thermal anomaly and have even investigated the degree to which the hot plume may exacerbate melting of the West Antarctic ice sheet.[35] But Antarctica's apparently stable position at the South Pole (due to wander) shouldn't necessarily be taken as evidence that plate tectonics doesn't have grand plans for the southern continent someday soon. Antarctica may be stable at the South Pole, but that doesn't necessarily mean it will always be tectonically stranded.

Geophysicist Roy Livermore of Cambridge also has his doubts that Antarctica would be left out of the next supercontinent. In Ted Nield's excellent book, *Supercontinent*, he interviewed Livermore, getting him to admit, "I don't believe Antarctica is going to stay at the pole." Livermore explained to Nield, "I want it to come north. Every other fragment of Gondwana has done that, piece by piece, and in the future Antarctica will; but only if it's dragged north by a subduction zone."[36] Livermore makes an excellent point. Earth's present megacontinent, Eurasia, has essentially grown to its current size because continental blocks once in the southern hemisphere (part of prior megacontinent Gondwana) transited across the equator to join Eurasia, the present megacontinent, in the northern hemisphere.[37] From one megacontinent to the next. Clearly these megacontinents—major, early building blocks of the larger, later supercontinents—are a major clue as to the dynamics of the supercontinent cycle that we are only just beginning to understand.[38]

Following up on Livermore's conjecture, the style of assembly of Earth's present megacontinent Eurasia may foreshadow the immediate steps in Earth's ongoing supercontinent cycle. As we mentioned before, a megacontinent like Eurasia appears to have formed with each past supercontinent cycle, typically about 200 million years before the larger supercontinent finally amalgamates. Eurasia is quite large already, but it also appears to not be done growing quite yet. My colleague at the Institute of Geology and Geophysics, Chinese Academy of Sciences (IGGCAS) in Beijing, Bo Wan, has researched in detail each and every suture between the many continental blocks that comprise the massive Eurasia. And he has found an obvious pattern.

Wan calls it the "one-way train" of subduction "that successively transferred the ruptured Gondwana continental fragments in the south, into the terminal in the north."[39] The most recent addition to Eurasia is the famous collision of India that gave rise to the towering Himalayas. As is always the case with colliding continents, an ocean had to close for such a collision to occur, and in the case of India's record-setting rapid plate tectonic motion, the collapse of two oceans may have been involved.[40] And this would have also been the case for the other continental blocks that came before India to join Eurasia (South China, the Lhasa block of Tibet, Indonesia, etc.). Each time an ocean would have had to close. The system of serial oceans, one after the other, is called the Tethys Ocean, with certain generations referred to, going forward through time, as the Proto-Tethys, the Paleo-Tethys, and the Neo-Tethys Oceans.

The present-day Indian Ocean is the current incarnation of the Tethys Ocean. Is there thus any reason to think the Indian Ocean would do any differently from all its ancestors before it? Livermore and Wan don't believe so. History is repeating. I would be remiss not to point out that closing the Indian Ocean, and thus adding Antarctica after Australia to Eurasia, would be entirely consistent with orthoversion. Like the Caribbean Sea and the Arctic Ocean, the Indian Ocean closing would allow Antarctica, at the South Pole 90° away from Africa and patiently waiting its turn, to someday join Amasia by migrating along the girdle. The only reason I leave my options open with respect to Antarctica is because the Scotia Sea between Antarctica and South America is also positioned along the girdle. In much the same way the Lesser Antilles arc can allow the Americas to rotate into each other, the same thing could very well happen between the Americas and Antarctica by way of the Scotia Sea closing. Like its underlying geology, hidden by its ice sheet, Antarctica's geodynamics also remain enigmatic.

Wan and I have yet to connect my orthoversion theory and his one-way-train theory in the scientific literature. I mention this unpublished idea of closing the Indian Ocean, and caution uncertainty with the competing option of closing the Scotia Sea, to make the point that our understanding of the supercontinent cycle is still a work in progress. We've certainly come a long way since Wegener, and I'm sure he is proud looking down. But we've still got a long way to go. With the framework laid out in this book—using the past as the key to the present, and even the future—I think it's not unreasonable to make an educated guess about the future. In my (admittedly biased) opinion, Amasia is our future, at least in the broad strokes of which oceans will close and why. This picture will invariably be adjusted in the coming years of continued research. But there's enough to go off already to warrant at least imaginative speculation of what the world of Amasia would be like. And it's impossible not to wonder if our distant descendants will be around to see the next supercontinent.

# EPILOGUE: SURVIVING AMASIA

Nothing in life is to be feared, it is only to be understood. Now is the
time to understand more, so that we may fear less.

MARIE CURIE

While the previous chapters have all offered the scientific research
that provides our current understanding of the supercontinent cycle,
I want to close this book with two philosophical questions. Could our
descendants live to see Amasia? And what exactly would they see?

In his book, *The Story of the Earth*, astrobiologist Robert Hazen
offers the provocative idea that the time for life on Earth may be limited. Hazen asserts that Earth only has five billion years left before
it is no longer in our solar system's "habitable zone," the Goldilocks
distance from the sun where it's not too cold or not too hot for liquid
water to exist.[1]

Unfortunately, I think worrying about the habitable zone 5 billion
years from now is overly optimistic. Hazen does not seem to consider
how plate tectonics—a critical ingredient for the cycling of nutrients
necessary for life—likely might not operate forever. Internal heat is
the fuel on which plate tectonics runs, of which Earth has only a finite
supply. Both Mars and the Moon once had such reserves of internal
heat, but billions of years of mantle convection burned through them
and they are now both tectonically dormant. Venus, a similar size to

Earth, may have had plate tectonics relatively recently in its past, but now convection in its mantle only occurs beneath a rigid, stagnant, immobile lid.[2] Thus Earth is the only body in the solar system that presently has plate tectonics. But this can't last forever. Even optimistic estimates for the longevity of plate tectonics—several billion more years—would present a problem to life on Earth well before our orbit has left the habitable zone.

I also can't help but imagine the challenge that forming another supercontinent only some 200 million years or so from now—admittedly still far off, but much sooner than billions of years from now—will present to our way of life. The next supercontinent would certainly threaten the coastal harbor cities that have been the lifeblood of our civilization. Amasia may not threaten the very existence of life on Earth. Lesser forms of life like bacteria that have survived many mass extinctions before will likely go on living. But Amasia will surely disrupt our way of life, no matter how technologically advanced we have become by then. And not only might humans not survive to witness Amasia, much of the rest of the animal kingdom may also face extinction. The Permian-Triassic extinction—the most lethal mass extinction in our planet's history—occurred the last time a supercontinent (Pangea) took shape. Amasia could be just as bad, or even worse.

•   •   •

But humans will have to clear many more immediate hurdles in order to bear witness to Amasia. When I started outlining this book several years ago, the threat of nuclear war seemed like the most imminent threat to humanity. But the end of 2019 showed our vulnerability to global pandemic. It's clear viral pandemics can prove an existential threat if unchecked. Luckily, modern medicine, social awareness, and hygiene have all improved substantially, and one can see the evidence of all this by comparing these past two most lethal cases. The 1918 flu infected about 500 million people, roughly one-third of the global population at the time, killing over 50 million people, whereas COVID-19 has to date infected 569 million people (only 7% of the current global population) and killed 6.4 million.[3] Whether a more lethal or adaptable novel virus will come along in the centuries to come is clearly something we must be vigilant about. As I finish edits of this manuscript, war in Europe makes the threat of nuclear war seem even more acute. Nuclear superpowers—having the most to lose with their large and growing economies—thus arguably must find other means to compete than by continuing to amass stockpiles of warheads that they will (hopefully) never use.

Assuming humans avoid both deadly pandemics and a nuclear holocaust, there is another self-inflicted challenge we must overcome: climate change. It is clear that climate change poses a real threat to our way of life, and that threat is disproportionately large for those that have the fewest resources to adapt. For example, rising sea levels disproportionally impact coastal countries that reply on lowland agriculture. Quite literally, entire economies and vast portions of the global food supply might go underwater. Yet, as with vaccines for pandemics and peace treaties for avoiding war, proposed solutions make me hopeful.

Some novel "geoengineering" solutions to global warming try to mimic plate tectonics. Take for instance the addition of aerosols to the atmosphere.[4] Aerosols increase Earth's albedo, thus reflecting more sunlight back into space and thereby cooling the planet. In 1991, the planet saw the effect of this after an eruption of Mount Pinatubo in the Philippines ejected millions of tons of sulfur dioxide into the atmosphere and lowered global temperatures by about 0.5°C from 1991 to 1993. At another scale entirely, a truly massive volcanic eruption occurring on the equator 720 million years ago may have injected enough aerosols to cause the global deep freeze of Snowball Earth.[5] These examples show both the effectiveness of atmospheric aerosol injection for abating global warming and the danger in going too far.

Planting more trees is certainly a good start, at least now. Currently, about 25% of carbon emissions from the use of fossil fuels is being taken up and stored by plants.[6] But in the long run, even a planting spree won't solve the problem. Indeed, trees fortunately consume carbon dioxide during photosynthesis, but they also leak some of that out the back end during respiration. To make matters worse, recent research reveals that as the world warms, plants will respire more, lessening over time the positive contribution made by plants.[7]

Another solution offered to counter some of the effects of climate change is carbon capture and sequestration—which is particularly important as countries still industrializing are unlikely to give up coal any time soon.[8] Both the capturing and the sequestering of carbon dioxide is straight out of the plate tectonic playbook. Subduction zones take hydrated slabs of oceanic crust back down to the mantle, and carbon dioxide is another gassy volatile in the rocky pore spaces that goes down along with all the water. If Earth can do it, why can't we? Carbon sequestration seeks to prevent carbon dioxide from entering the atmosphere by storing it in geological rock formations for long periods of time. One option is burying dead trees. The carbon stored in them will stay locked away longer than if we let them decompose.[9] Other examples even include coal-fired power plants that can catch the carbon dioxide they release and inject it back into porous rock formations underlying an impermeable cap layer. The lingering question about carbon capture is whether humans can do it on a scale to offset carbon emissions.

Such bold geoengineering actions, coupled with equally bold conservation of resources, might help us address the threats of climate change in the decades and centuries ahead. But one must also admit the limitations of such solutions. For example, you can lower global temperature, but you won't stop ocean acidification by adding aerosols to the atmosphere. Also, to be clear, solving global warming isn't about "saving the planet." It's about saving *us*. Many animals and plants have survived, even thrived, under much more severe greenhouse climates in Earth's past. But we haven't. Earth will be fine without us. But we won't be fine without an Earth that works for us.

•   •   •

Assuming we are able to reverse the exponential warming trend started at the dawn of the Industrial Revolution, humans would presumably next be presented with the ice age that global warming has temporarily delayed. The ice ages of recent climate history are caused by the Milankovitch cycles, named for Milutin Milankovitch, a Serbian geophysicist and astronomer. In the 1920s he theorized that changes in Earth's orbit and rotation could influence how much incoming solar radiation energy reaches Earth's surface, thereby controlling climate. The Milankovitch cycles that explain the pacing of the repeated ice ages include the precession (or wobble) and the obliquity (or tilt) of Earth's spin axis in space with cycles of about 20,000 and 41,000 years, respectively, as well as the eccentricity (versus circularity) of Earth's orbit with a cycle of about 100,000 years. This last factor has been particularly dominant in the pacing of the most severe ice ages for the past 1 million years since the mid-Pleistocene.

But perhaps the most fascinating thing about viewing an impending ice age as a threat is its deep irony. After all, it was during the most recent ice age that *Homo sapiens* as a species took hold. Before the most recent interglacial Holocene epoch about 12,000 years ago, the peak of the ice age occurred about 18,000 years ago, which can be explained by the pacing of the precession cycle. Our new species was uniquely capable of adapting to the harsh conditions in innovative ways, such as by sewing warm clothing with bone needles and migrating between continents across land bridges exposed by low sea levels. Such ice ages will surely come back again and we can only hope we will fare as well as our forebearers. Our ancestors survived, and even thrived, in earlier such periods, but surely our way of life would be strained and changed even if our species escapes extinction.

Assuming we were able to survive the first severe ice age, we would then have to find a way to live through the ebb and flow of the natural glacial-interglacial cycles that would change our environment between extremes. Over the past 10 million years, only a handful of species of mammals have lasted for more than 6 million years. Most have survived fewer than 4 million years, and the average longevity of mammal species is 3.4 million years.[10] Embracing human exceptionalism, and for the sake of speculation, let's just assume humans or a related species can beat the odds.

Assuming we found a way to live sustainably through multiple ice age cycles, and that no asteroids collided with Earth (which tends to happen about every 20 million years), we'd then have to adapt to the climate changes that the assembly of Amasia itself would necessarily bring. An ice age with Amasia centered on the North Pole, however, would be much more severe than the recent ice ages of our ancestors. The last time Earth formed such a large landmass at high latitudes was the formation of megacontinent Gondwana at the South Pole during the Paleozoic era. Aside from globe-girdling Snowball Earth, the Gondwanan ice sheet was the largest in Earth history, and it covered most of the megacontinent at its peak.[11] Amasia, a full-fledged supercontinent, would have an even larger ice sheet than Gondwana hosted.

Even if we found ways to survive around the edges of the Amasian ice sheet, life is just harder at high latitudes. The latitudinal diversity gradient is perhaps the best-known feature of diversity on Earth:

there are more species in the warm tropics around the equator and fewer at cold, high latitudes.[12] It turns out life is hard at high latitudes, where very regular species turnover occurs.[13] It's likely that Africa, the best candidate for preserving some equatorial ecospace on Amasia, would be our best option for survival (fig. 50). We'd have to come full circle, back to the cradle of humankind.

There is always the chance that wander would save us from a polar supercontinent. The mantle upwelling developing beneath a newly formed supercontinent would like to be on the equator. Hosting the largest volcano in the solar system, Olympus Mons, the Tharsis volcanic province on Mars is also the largest single mass on a celestial body in the solar system, and it sits squarely on the equator. This is not by coincidence; wander brought it there.[14] We can thus hope that if plate tectonics attempts once again, as it did with Gondwana, to form Amasia at the pole, wander may be our savior by shifting the whole globe to put Amasia and its underlying mantle upwelling snuggly on the equator. After all, all of the past three supercontinents, once fully formed, have straddled the equator (fig. 38). Once again, this is not coincidence but geophysics.

As we have learned very well from the past, however, even equatorial supercontinents can cause trouble. Massive climatic, environmental, and ecological changes would ensue under a supercontinent world again. Given that the Permian-Triassic extinction during Pangea obliterated well over half of living species (fig. 12), will we be able to survive Amasia? Species richness around us would likely plummet, just like it did before. The many isolated islands of continents harboring well-adapted species akin to Darwin's finches would vanish in the ecological sameness of a supercontinent. Land bridges would form, allowing migrations of successful species into distant lands, introducing new competition that once well-adapted species (like the dodo bird) didn't used to have to deal with and may not be able to sustain. But Amasia would also form some natural ecological barriers of its own in the form of new great mountain chains, such as those that divided Pangea into hemispheres supporting their different species of dinosaurs. But back to us and our survival.

Finally, our beloved coastlines will collide as Amasia assembles. One of the more imminent tectonic catastrophes would likely be the turning of the Mediterranean Sea into the Mediterranean Mountains. Much of populated Europe and northern Africa would have to pick up and migrate elsewhere. The cradles of ancient civilizations in Egypt and Greece would be all but obliterated, either thrust up high in the sky like a new chain of Alps or shoved down deep to the great pressures and temperatures of the Earth's crust.

In the introduction, I might have teased you with the evocative image of skyscrapers needling their way down intact into the gaping jaws of subduction zones. But reality will likely be less Hollywood. The increasing number of earthquakes in cities like New York, once along a passive margin that would become active tectonically, would likely be enough to raze skyscrapers long before continental collisions forced them down subduction zones. Like the geologic mélange of debris that gets wedged into subduction zones, the remains of our once great cities would just be rubble added to the downgoing pile by that point.

But these new continental collisions would threaten modern civilization not only with structural damages. Possibly the greatest threat the shifting plates would create would be the sheer scale of human and cultural displacement. The most devastating example I can think of would be the closure of the Caribbean Sea. According to our very rough guesstimate, Lima, Peru, on the west coast of South America, and New York City, on the eastern seaboard of the United States, are on a collision course: New York City being the most populated city of one of the world's great superpowers (presently just shy of 9 million residents) and a major hub of the global economy; Lima being one of the most populated cities in all the Americas with over 9 million current residents. Major population centers all around the world would have to relocate either inland or to safer shores. Antarctica, in the event it doesn't stay stranded at the South Pole and migrates north, may change from being a land of cold desolation to our last, best hope.

•    •    •

From this thought experiment about the future, it may seem like the clock is running out on our species, which not only reduces our chance of living on the next supercontinent but also for *even seeing* the next supercontinent modeled. It should be clear by now that it takes time for humanity to gather evidence and wrestle with ideas. We will someday, hopefully soon, have an even better idea of how Amasia will take shape. The only question is whether we can give ourselves enough time to make such discoveries. It takes generations to solve big problems. Can society learn to make decisions like Native Americans did, thinking how their actions would affect seven generations hence? As our story started with Alfred Wegener and ends with my own ongoing humble efforts, it is thus not hyperbole to assert that fundamental problems in science as large as how supercontinents form are only solved by multiple generations carrying the torch forward.

I use the torch metaphor for several reasons. First, it was the way former president Obama viewed his time in one of the most powerful jobs in the world. Despite the wieldy power, despite the desperate, crisis-ridden undercurrent of the times, he knew he was just doing the best he could in his moment in time before passing the torch on. He was only a small part of a long and ongoing legacy of efforts to perfect our union. I also use this analogy because fire is one of humanity's earliest significant innovations. The innovations explored in this book are twofold: developing the scientific method and then turning it on the Earth to understand the planet we call home. Humanity's new torch is passing on the fire of the scientific method to the next generation: Can they answer the questions that have eluded us? The good news is they have the tools; the bad news is that they may not have the time.

In my young career, I have already encountered and been encouraged by the generations to come, those to soon inherit the Earth. My students are as hungry as I once was—but they also have more of a built-in awareness of the world around them. When I was growing up, overpopulation was really the only existential problem facing humanity of which we were widely aware. Environmentalism was certainly beginning to get its foothold, but it was only loosely linked to other larger looming issues such as global warming. Only by the time that I was in university had Al Gore made his mark not only in the hushed halls of Capitol Hill but also on the silver screen. The real inconvenient truth is that the Industrial Revolution has given us everything we value as achievements of modern society: high quality of life (heating, lighting, motorized transportation, modern medicine, etc.), increased life expectancy (and thus also grandparenting), and even walking on the Moon. Sadly, the side effects of the fossil fuel era have been widespread and significant, and they can now be ignored only at our peril.

Can we buy enough time to solve the problems facing us? Theories take time to develop and they take even more time to test. When I published my orthoversion theory in *Nature*, Paul Hoffman apparently gave me, through a colleague of mine, a compliment that he knew would somehow find its way back to me (he would never be so effusive in person). He had apparently said something to the effect that our linking the supercontinent cycle to mantle convection represented "the biggest advance in the field of supercontinents in decades." This was 2012, roughly twenty years after his "inside out" paper published in *Science* that really got geologists out on the hunt for ancient supercontinents. And it's now been a decade since my paper. There has surely been progress since then, but I am under no illusion that the theory has been adequately tested—it certainly has yet to be disproven, but neither can I claim that it has been validated by any sense of consensus yet.

Science takes time. This is both at once frustrating and a redeeming quality. Science is now a massive, globe-girdling, interconnected web of a community—a small network that has become vast in scale. And there is certainly also a "science economy" that has an inertia of its own. Just like Michael Bay's increasing pyrotechnics budget from one *Transformers* movie to the next, according to such incremental, capitalist measures of success, scientists viewed as successful today are those attracting the largest grants. Is increased interest in space and planetary science fueled by the curiosity of human exploration or a thirst for billion-dollar budgets?

Luckily, despite these jaded realities that shape modern science, each new generation is born with an unvarnished curiosity. The elementary school students that have watched my simulations of Amasia on YouTube will be the ones carrying the torch in this story. Thank you for joining me on this time-traveling voyage from past to future worlds. Everything is not yet lost, but we do need time. Only with an eye on the future can we find a way to be a part of it.

# ACKNOWLEDGMENTS

I must foremost thank my colleagues and dear friends David Evans and Taylor Kilian—my coauthors on the 2012 *Nature* paper that brought supercontinent science back into the mainstream.

It's not easy writing your first book, but if you get the amazing amounts of help that I did, apparently it's possible. I must thank my editor, Joe Calamia, who, as embarrassing as it is for me to admit, came up with the idea for this book. I had originally approached Joe about writing a book about Alfred Wegener's prescience. Joe kindly declined but wisely suggested that I write about the next supercontinent—the one thing I was known for outside of academia. Joe's vision turned out to be just the beginning of his unparalleled contributions to this book. I might have had an iota of natural ability in this regard, but Joe taught me over months and years of numerous manuscript iterations how to write popular science—it certainly didn't hurt that his scientific mind is sharp, his positivity is infectious, and his personality is delightful. With certainly no disrespect to my alma mater Yale University and its own prestigious press, there was never any hesitation in my decision to have the project follow Joe to his new home at the University of Chicago Press.

Brendan Murphy and Richard Ernst both reviewed (anonymously at the time) both my book proposal as well as my submitted manuscript for Chicago. Richard brought the same enthusiastic and authentic critical eye that he brings to his research. Brendan's contributions are utterly incalculable. His experience writing very successful earth science textbooks allowed him to make copious edits that truly brought the book to the next level. It's no exaggeration of my appreciation to say that first and last rounds will forever be on me.

I must thank my teachers, but I have had so many great ones that I cannot mention them all. In fifth grade, I first learned about plate tectonics from Mrs. Charbeneau. In seventh grade, I went home every day with stories told by Mr. Arndt about the lives of great scientists and how they were inspired to make their great discoveries. I had countless influential teachers at Kent School. Mr. and Mrs. Goodwin made me a scientist and Charles Gould taught me to write—and to love doing it. In college, Cameron Davidson did one-on-one independent studies with me to supplement Carleton's curriculum in geology with developing interests of my own. Nicholas Swanson-Hysell provided me a role model and gave me confidence at the very beginning of my career in geology—I will never forget the night in the damp Dolomites when he gave me David Evans's paper on wander to read: I was completely overwhelmed and enthused at once. Alessandro Montanari taught me that nothing, even in the hallowed halls of published science, "nothing is certain." Sarah Mitchell (no relation) taught my first course in geology and made me realize that I was destined for it. David Bice taught me the art of writing a scientific paper.

Since graduate school, countless professors have patiently and passionately nurtured my abilities to conduct scientific research. Joe Kirschvink essentially taught me everything—most importantly that it's OK to like your hypothesis if you love testing it. David Evans not only guided me to make the right decision for graduate school, he also followed through with the "training in tectonics" he had promised as we listened to James Brown outside the Owl Shop in downtown New Haven. But beyond his first-rate mentorship of me, David inspired my mind with his research like no one else. Also at Yale, Mark Brandon always told me what I needed to hear. Wouter Bleeker effectively served as my second PhD advisor and taught me everything I know about how to do remote field work. Thanks to Paul Hoffman for being to me what Wegener was to him: the ultimate inspiration. Peter Ward single-handedly inspired me to aspire to a second career in popular science, not only with his amazing books like *Gorgon* and *Rare Earth*, but with a personal friendship for the ages. Zheng-Xiang Li gave me the opportunity of a lifetime to work with an interdisciplinary science team and put me in an office with Brendan Murphy himself. Peng Peng brought me to China, starting the next chapter of my research career that gave me the stability to finish this book. Everyone at the Institute of Geology and Geophysics, Chinese Academy of Sciences (IGGCAS), particularly past Director Fuyuan Wu, has provided me with unfailing support and exciting opportunities. I am greatly indebted to my students and to the energetic staff at IGGCAS. The collaborators across the globe that have influenced me are too many to name, but you know who you are.

My grandfather on my mother's side, Nelson "Poppa" Burrin, was one of a kind and left his fingerprints on my mind and character. Nelson was a chemical engineer during the plastics revolution. As Sinatra sang, Poppa taught me to stay "young at heart." Nelson's life and legacy inspire me every day. My cousins, aunts, and uncles have always made life fun and personal. My parents, James and Suzi, met as English teachers and painstakingly developed my writing. My father, a writer, and my mother, an artist, also taught me to love the creative process. They are the best parents imaginable. My older brother Nicholas gave me a role model from day one and has never wavered from that role. Our "constructive competition"—which we both acknowledged on a stroll through Les Invalides—pushes me to this day. My sister-in-law, Jocelyn, is the sister I never had and has guided me brilliantly through the twists and turns of academia. Their two sons, my dear nephews Graham and Boden, are the lights of our lives. Friends mean so much to me and you know who you are. Taylor and Becca got me through graduate school. Charlie Nathan has shared a parallel universe with me, from academia to beat-making. Grant Cox, Luc Doucet, Thomas Gernon, Uwe Kirscher, Erin Martin, Adam Nordsvan, and Chris Spencer made for a great time down under. Xiaofang He, Peng Peng, Bo Wan, and Lei Zhao have made Beijing feel like home. Sam and Vince and their lovely families give my life meaning and make it sing.

Many thanks to Harry Netzer, Selena Pang, and Chris Spencer for reading portions of the proposal and manuscript along the long journey. My copyeditor, and awesomely incidentally a fellow Carl, Susan Olin, made me enjoy reading my book much more than before. Working with Matthew Green as he artistically imagined the captivating chapter opening illustrations was a fun and wild ride, and his commitment to portraying Earth history was appreciated. Sanjiv Kumar Sinha prepared a beautiful index on a very tight schedule. Many of the figures involving paleogeographic reconstructions were made with the wonderful GPlates software. Above all, thanks to the University of Chicago Press for taking a chance on a first-time author. Let's do it again sometime.

# NOTES

### CHAPTER ONE

1. James Brendan Murphy and R. Damian Nance, "The Pangea conundrum," *Geology* 36, no. 9 (2008): 703–6.

2. Alfred Wegener, "Die Entstehung der Kontinente," *Geologische Rundschau* 3, no. 4 (1912): 276–92; Alfred Wegener, "Die Entstehung der Kontinente und Ozeane: Braunschweig," *Sammlung Vieweg* 23 (1915): 94; Alfred Wegener, *Die Entstehung der Kontinente und Ozeane*, vol. 66 (F. Vieweg, 1920); A. Wegener, *The origin of continents and oceans*, trans. J. Biram (English translation of the 4th [1929] German edition) (London: Dover, 1966).

3. Mott T. Greene, *Alfred Wegener: Science, exploration, and the theory of continental drift* (Baltimore: Johns Hopkins University Press, 2015).

4. Alexander Logie Du Toit, *Our wandering continents: An hypothesis of continental drifting* (London: Oliver and Boyd, 1937).

5. See https://www.amnh.org/learn-teach/curriculum-collections/dinosaurs -activities-and-lesson-plans/plate-tectonics-puzzle.

6. Ralph Dietmar Müller, Maria Sdrolias, Carmen Gaina, and Walter R. Roest, "Age, spreading rates, and spreading asymmetry of the world's ocean crust," *Geochemistry, Geophysics, Geosystems* 9, no. 4 (2008).

7. Ronald E. Doel, Tanya J. Levin, and Mason K. Marker, "Extending modern cartography to the ocean depths: Military patronage, Cold War priorities, and the Heezen-Tharp mapping project, 1952–1959," *Journal of Historical Geography* 32, no. 3 (2006): 605–26.

8. Bruce C. Heezen and Marie Tharp, "Tectonic fabric of the Atlantic and Indian Oceans and continental drift," *Philosophical Transactions of the Royal Society of London, Series A, Mathematical and Physical Sciences* 258, no. 1088 (1965): 90–106; Bruce C. Heezen, Marie Tharp, and Maurice Ewing, *The floors of the oceans*, vol. 65 (New York: Geological Society of America, 1959).

9.  Ott Christoph Hilgenberg, "Vom wachsenden Erdball," *Charlottenburg* (1933).

10. William Thomson, "XV.—On the secular cooling of the Earth." *Earth and Environmental Science Transactions of the Royal Society of Edinburgh* 23, no. 1 (1862): 157–69.

11. John Tuzo Wilson, "A new class of faults and their bearing on continental drift," *Nature* 207, no. 4995 (1965): 343–47.

12. Wegener, *Origin of continents and oceans.*

13. Fred J. Vine and D. H. Matthews, "Magnetic anomalies over oceanic ridges," *Nature* 199, no. 4897 (1963): 947–49; F. J. Vine, "Spreading of the ocean floor: New evidence: Magnetic anomalies may record histories of the ocean basins and Earth's magnetic field for 2 × 108 years," *Science* 154, no. 3755 (1966): 1405–15.

14. Tinsley H. Davis, "Biography of Edward Irving," *Proceedings of the National Academy of Sciences* 102, no. 6 (2005): 1819–20.

15. Edward Irving and R. Green, "Polar movement relative to Australia," *Geophysical Journal International* 1, no. 1 (1958): 64–72.

16. Edward Irving, "Drift of the major continental blocks since the Devonian," *Nature* 270, no. 5635 (1977): 304–9.

17. John Tuzo Wilson, "Evidence from islands on the spreading of ocean floors," *Nature* 197, no. 4867 (1963): 536–38.

18. John Tuzo Wilson, "Did the Atlantic close and then re-open?," *Nature* 211, no. 5050 (1966): 676–81.

19. Wilson, "Did the Atlantic close and then re-open?," 676.

20. John Tuzo Wilson, "Static or mobile earth: The current scientific revolution," *Proceedings of the American Philosophical Society* 112, no. 5 (1968): 309–20.

21. Trond H. Torsvik, Rob Van der Voo, Ulla Preeden, et al., "Phanerozoic polar wander, palaeogeography and dynamics," *Earth-Science Reviews* 114, no. 3–4 (2012): 325–68.

22. Charles R. Marshall, "Explaining the Cambrian 'explosion' of animals," *Annual Review of Earth and Planetary Sciences* 34 (2006): 355–84.

23. Peter D. Ward, Jennifer Botha, Roger Buick, et al., "Abrupt and gradual extinction among Late Permian land vertebrates in the Karoo Basin, South Africa," *Science* 307, no. 5710 (2005): 709–14.

24. Andrew Zaffos, Seth Finnegan, and Shanan E. Peters, "Plate tectonic regulation of global marine animal diversity," *Proceedings of the National Academy of Sciences* 114, no. 22 (2017): 5653–58.

25. Edward J. Garnero, Thorne Lay, and Allen McNamara, "Implications of lower-mantle structural heterogeneity for the existence and nature of

whole-mantle plumes," *Special Papers—Geological Society of America* 430 (2007): 79.

26. Ross N. Mitchell, Lei Wu, J. Brendan Murphy, and Zheng-Xiang Li, "Trial by fire: Testing the paleolongitude of Pangea of competing reference frames with the African LLSVP," *Geoscience Frontiers* 11, no. 4 (2020): 1253–56.

27. Murphy and Nance, "Pangea conundrum."

CHAPTER TWO

1. Timothy D. Raub, J. L. Kirschvink, and D. A. D. Evans, "True polar wander: Linking deep and shallow geodynamics to hydro-and bio-spheric hypotheses," *Treatise on Geophysics* 5 (2007): 565–89.

2. Paul F. Hoffman, "Did the breakout of Laurentia turn Gondwanaland inside-out?," *Science* 252, no. 5011 (1991): 1409–12.

3. Eldridge M. Moores, "Southwest US-East Antarctic (SWEAT) connection: A hypothesis," *Geology* 19, no. 5 (1991): 425–28; Ian W. D. Dalziel, "Pacific margins of Laurentia and East Antarctica-Australia as a conjugate rift pair: Evidence and implications for an Eocambrian supercontinent," *Geology* 19, no. 6 (1991): 598–601.

4. Hoffman, "Did the breakout of Laurentia turn Gondwanaland inside-out?"

5. Erin L. Martin, C. J. Spencer, W. J. Collins, R. J. Thomas, P. H. Macey, and N. M. W. Roberts, "The core of Rodinia formed by the juxtaposition of opposed retreating and advancing accretionary orogens," *Earth-Science Reviews* 211 (2020): 103413.

6. Zheng-Xiang Li, S. Bogdanova, A. S. Collins, et al., "Assembly, configuration, and break-up history of Rodinia: A synthesis," *Precambrian Research* 160, no. 1–2 (2008): 179–210.

7. Li, Bogdanova, Collins, et al., "Assembly, configuration, and break-up history of Rodinia."

8. Zheng-Xiang Li and David A. D. Evans, "Late Neoproterozoic 40 intraplate rotation within Australia allows for a tighter-fitting and longer-lasting Rodinia," *Geology* 39, no. 1 (2011): 39–42.

9. William J. Collins, J. Brendan Murphy, Tim E. Johnson, and Hui-Qing Huang, "Critical role of water in the formation of continental crust," *Nature Geoscience* 13, no. 5 (2020): 331–38.

10. Martin, Spencer, Collins, Thomas, Macey, and Roberts, "Core of Rodinia."

11. Allen P. Nutman, Vickie C. Bennett, Clark R. L. Friend, Martin J. Van Kranendonk, and Allan R. Chivas. "Rapid emergence of life shown by discovery of 3,700-million-year-old microbial structures," *Nature* 537, no. 7621 (2016): 535–38.

12. Nicholas L. Swanson-Hysell, Adam C. Maloof, Joseph L. Kirschvink,

David A. D. Evans, Galen P. Halverson, and Matthew T. Hurtgen, "Constraints on Neoproterozoic paleogeography and Paleozoic orogenesis from paleomagnetic records of the Bitter Springs Formation, Amadeus Basin, central Australia," *American Journal of Science* 312, no. 8 (2012): 817–84.

13. Clinton P. Conrad and Bradford H. Hager, "Mantle convection with strong subduction zones," *Geophysical Journal International* 144, no. 2 (2001): 271–88.

14. Adam C. Maloof, Galen P. Halverson, Joseph L. Kirschvink, Daniel P. Schrag, Benjamin P. Weiss, and Paul F. Hoffman, "Combined paleomagnetic, isotopic, and stratigraphic evidence for true polar wander from the Neoproterozoic Akademikerbreen Group, Svalbard, Norway," *Geological Society of America Bulletin* 118, no. 9-10 (2006): 1099–1124.

15. Maloof, Halverson, Kirschvink, Schrag, Weiss, and Hoffman, "Combined paleomagnetic, isotopic, and stratigraphic evidence."

16. Swanson-Hysell, Maloof, Kirschvink, Evans, Halverson, and Hurtgen, "Constraints on Neoproterozoic paleogeography and Paleozoic orogenesis."

17. Robert H. Rainbird, Larry M. Hearnan, and Grant Young. "Sampling Laurentia: Detrital zircon geochronology offers evidence for an extensive Neoproterozoic river system originating from the Grenville orogen," *Geology* 20, no. 4 (1992): 351–54.

18. Xianqing Jing, Zhenyu Yang, David A. D. Evans, Yabo Tong, Yingchao Xu, and Heng Wang, "A pan-latitudinal Rodinia in the Tonian true polar wander frame," *Earth and Planetary Science Letters* 530 (2020): 115880.

19. Bernhard Steinberger, Miriam-Lisanne Seidel, and Trond H. Torsvik, "Limited true polar wander as evidence that Earth's nonhydrostatic shape is persistently triaxial," *Geophysical Research Letters* 44, no. 2 (2017): 827–34.

20. Jessica R. Creveling, J. X. Mitrovica, N.-H. Chan, K. Latychev, and I. Matsuyama, "Mechanisms for oscillatory true polar wander," *Nature* 491, no. 7423 (2012): 244–48.

## CHAPTER THREE

1. Alexander Logie Du Toit, *Our wandering continents: An hypothesis of continental drifting* (London: Oliver and Boyd, 1937).

2. Sergei A. Pisarevsky and L. M. Natapov, "Siberia and Rodinia," *Tectonophysics* 375, no. 1-4 (2003): 221–45; David A. D. Evans and Ross N. Mitchell, "Assembly and breakup of the core of Paleoproterozoic-Mesoproterozoic supercontinent Nuna," *Geology* 39, no. 5 (2011): 443–46.

3. Guochun Zhao, Peter A. Cawood, Simon A. Wilde, and Min Sun, "Review of global 2.1-1.8 Ga orogens: Implications for a pre-Rodinia supercontinent," *Earth-Science Reviews* 59, no. 1-4 (2002): 125–62.

4.   Paul F. Hoffman, "Tectonic genealogy of North America," *Earth structure: An introduction to structural geology and tectonics* (1997): 459–64.

5.   John J. W. Rogers and M. Santosh, "Configuration of Columbia, a Mesoproterozoic supercontinent," *Gondwana Research* 5, no. 1 (2002): 5–22.

6.   Zhao, Cawood, Wilde, and Sun, "Review of global 2.1–1.8 Ga orogens."

7.   Gabrielle Walker, *Snowball Earth: The story of a maverick scientist and his theory of the global catastrophe that spawned life as we know it* (London: Bloomsbury, 2003).

8.   Zhao, Cawood, Wilde, and Sun, "Review of global 2.1–1.8 Ga orogens."

9.   Glen Arthur Izett and Ray Everett Wilcox, *Map showing localities and inferred distributions of the Huckleberry Ridge, Mesa Falls, and Lava Creek ash beds (Pearlette family ash beds) of Pliocene and Pleistocene age in the western United States and southern Canada*, no. 1325, 1982; http://pubs.er.usgs.gov /publication/i1325 IMAP series, Report 1325, doi: 10.3133/i1325.

10.  Paul Hoffman, "Proterozoic paleocurrents and depositional history of the East Arm fold belt, Great Slave Lake, Northwest Territories," *Canadian Journal of Earth Sciences* 6, no. 3 (1969): 441–62.

11.  Samuel A. Bowring, W. R. Van Schmus, and P. F. Hoffman, "U–Pb zircon ages from Athapuscow aulacogen, east arm of Great Slave Lake, NWT, Canada," *Canadian Journal of Earth Sciences* 21, no. 11 (1984): 1315–24.

12.  Adam R. Nordsvan, William J. Collins, Zheng-Xiang Li, et al., "Laurentian crust in northeast Australia: Implications for the assembly of the supercontinent Nuna," *Geology* 46, no. 3 (2018): 251–54.

13.  David A. D. Evans and Ross N. Mitchell, "Assembly and breakup of the core of Paleoproterozoic–Mesoproterozoic supercontinent Nuna," *Geology* 39, no. 5 (2011): 443–46.

14.  Amaury Pourteau, Matthijs A. Smit, Zheng-Xiang Li, et al., "1.6 Ga crustal thickening along the final Nuna suture," *Geology* 46, no. 11 (2018): 959–62.

15.  Chong Wang, Ross N. Mitchell, J. Brendan Murphy, Peng Peng, and Christopher J. Spencer, "The role of megacontinents in the supercontinent cycle," *Geology* 49, no. 4 (2021): 402–6.

16.  Paul F. Hoffman, "Speculations on Laurentia's first gigayear (2.0 to 1.0 Ga)," *Geology* 17, no. 2 (1989): 135–38.

17.  Hoffman, "Speculations on Laurentia's first gigayear."

18.  Hoffman, "Speculations on Laurentia's first gigayear."

19.  Don L. Anderson, Hotspots, polar wander, Mesozoic convection and the geoid," *Nature* 297, no. 5865 (1982): 391–93.

20.  Jun Korenaga, "Eustasy, supercontinental insulation, and the temporal variability of terrestrial heat flux," *Earth and Planetary Science Letters* 257, no. 1–2 (2007): 350–58.

21.  Korenaga, "Eustasy, supercontinental insulation."

22. Shijie Zhong, Nan Zhang, Zheng-Xiang Li, and James H. Roberts, "Super-continent cycles, true polar wander, and very long-wavelength mantle convection," *Earth and Planetary Science Letters* 261, no. 3-4 (2007): 551–64.

23. Zhong, Zhang, Li, and Roberts, "Supercontinent cycles."

24. Hoffman, "Speculations on Laurentia's first gigayear."

## CHAPTER FOUR

1. Alexei L. Perchuk, Taras V. Gerya, Vladimir S. Zakharov, and William L. Griffin, "Building cratonic keels in Precambrian plate tectonics," *Nature* 586, no. 7829 (2020): 395–401.

2. R. Roberts, "Ore deposit models #11. Archean lode gold deposits," *Geoscience Canada* 14, no. 1 (1987): 37–52.

3. David I. Groves, G. Neil Phillips, Susan E. Ho, Sarah M. Houstoun, and Christine A. Standing, "Craton-scale distribution of Archean greenstone gold deposits; predictive capacity of the metamorphic model," *Economic Geology* 82, no. 8 (1987): 2045–58.

4. Nicholas Arndt, Michael Lesher, and Steve Barnes, *Komatiite* (New York: Cambridge University Press, 2008).

5. Stephen J. Barnes and N. T. Arndt, "Distribution and geochemistry of komatiites and basalts through the Archean," in *Earth's Oldest Rocks*, ed. Martin J. Van Kranendonk, Vickie C. Bennett, and J. Elis Hoffmann (Elsevier, 2019).

6. Reid R. Keays, "The role of komatiitic and picritic magmatism and S-saturation in the formation of ore deposits," *Lithos* 34, no. 1-3 (1995): 1–18.

7. Heinrich D. Holland, "Volcanic gases, black smokers, and the Great Oxidation Event," *Geochimica et Cosmochimica Acta* 66, no. 21 (2002): 3811–26.

8. Andrew H. Knoll and Martin A. Nowak, "The timetable of evolution," *Science Advances* 3, no. 5 (2017): e1603076; Allen P. Nutman, Vickie C. Bennett, Clark R. L. Friend, Martin J. Van Kranendonk, and Allan R. Chivas, "Rapid emergence of life shown by discovery of 3,700-million-year-old microbial structures," *Nature* 537, no. 7621 (2016): 535–38.

9. Thomas A. Laakso and Daniel P. Schrag, "Regulation of atmospheric oxygen during the Proterozoic," *Earth and Planetary Science Letters* 388 (2014): 81–91; Grant M. Cox, Timothy W. Lyons, Ross N. Mitchell, Derrick Hasterok, and Matthew Gard, "Linking the rise of atmospheric oxygen to growth in the continental phosphorus inventory," *Earth and Planetary Science Letters* 489 (2018): 28–36.

10. Toby Tyrrell, "The relative influences of nitrogen and phosphorus on oceanic primary production," *Nature* 400, no. 6744 (1999): 525–31.

11. Cin-Ty A. Lee, Laurence Y. Yeung, N. Ryan McKenzie, Yusuke Yokoyama,

Kazumi Ozaki, and Adrian Lenardic, "Two-step rise of atmospheric oxygen linked to the growth of continents," *Nature Geoscience* 9, no. 6 (2016): 417–24; I. N. Bindeman, D. O. Zakharov, J. Palandri, et al., "Rapid emergence of subaerial landmasses and onset of a modern hydrologic cycle 2.5 billion years ago," *Nature* 557, no. 7706 (2018): 545–48.

12. Alan M. Goodwin, *Principles of Precambrian geology* (London: Elsevier, 1996).

13. Meng Guo and Jun Korenaga, "Argon constraints on the early growth of felsic continental crust," *Science Advances* 6, no. 21 (2020): eaaz6234.

14. Jun Korenaga, "Crustal evolution and mantle dynamics through Earth history," *Philosophical Transactions of the Royal Society A: Mathematical, Physical and Engineering Sciences* 376, no. 2132 (2018): 20170408.

15. Ross N. Mitchell, Nan Zhang, Johanna Salminen, et al., "The supercontinent cycle," *Nature Reviews Earth & Environment* 2, no. 5 (2021): 358–74.

16. Bindeman, Zakharov, Palandri, et al., "Rapid emergence of subaerial landmasses."

17. Norman H. Sleep, "Martian plate tectonics," *Journal of Geophysical Research: Planets* 99, no. E3 (1994): 5639–55; Jafar Arkani-Hamed, "On the tectonics of Venus," *Physics of the Earth and Planetary Interiors* 76, no. 1-2 (1993): 75–96.

18. Allen P. Nutman, "Antiquity of the oceans and continents," *Elements* 2, no. 4 (2006): 223–27; Timothy Kusky, Brian F. Windley, Ali Polat, Lu Wang, Wenbin Ning, and Yating Zhong, "Archean dome-and-basin style structures form during growth and death of intraoceanic and continental margin arcs in accretionary orogens," *Earth-Science Reviews* 220 (2021): 103725.

19. Tim E. Johnson, Michael Brown, Nicholas J. Gardiner, Christopher L. Kirkland, and R. Hugh Smithies, "Earth's first stable continents did not form by subduction," *Nature* 543, no. 7644 (2017): 239–42; Jean H. Bédard, "A catalytic delamination-driven model for coupled genesis of Archaean crust and sub-continental lithospheric mantle," *Geochimica et Cosmochimica Acta* 70, no. 5 (2006): 1188–1214.

20. Samuel A. Bowring, I. S. Williams, and W. Compston, "3.96 Ga gneisses from the slave province, Northwest Territories, Canada," *Geology* 17, no. 11 (1989): 971–75; Samuel A. Bowring and Ian S. Williams, "Priscoan (4.00–4.03 Ga) orthogneisses from northwestern Canada," *Contributions to Mineralogy and Petrology* 134, no. 1 (1999): 3–16.

21. William Compston and Robert T. Pidgeon, "Jack Hills, evidence of more very old detrital zircons in Western Australia," *Nature* 321, no. 6072 (1986): 766–69.

22. Timothy Mark Harrison, "The Hadean crust: Evidence from > 4 Ga zircons," *Annual Review of Earth and Planetary Sciences* 37 (2009): 479–505.

23.  C. Brenhin Keller, Patrick Boehnke, and Blair Schoene, "Temporal varia-
     tion in relative zircon abundance throughout Earth history," *Geochemical
     Perspectives Letters* 3 (2017): 179–89.

24.  Simon A. Wilde, John W. Valley, William H. Peck, and Colin M. Graham,
     "Evidence from detrital zircons for the existence of continental crust and
     oceans on the Earth 4.4 Gyr ago," *Nature* 409, no. 6817 (2001): 175–78.

25.  Bo Wan, Xusong Yang, Xiaobo Tian, Huaiyu Yuan, Uwe Kirscher, and Ross
     N. Mitchell, "Seismological evidence for the earliest global subduction net-
     work at 2 Ga ago," *Science Advances* 6, no. 32 (2020): eabc5491.

26.  Wouter Bleeker, "The late Archean record: A puzzle in ca. 35 pieces," *Lithos*
     71, no. 2–4 (2003): 99–134.

27.  Ross N. Mitchell, Wouter Bleeker, Otto Van Breemen, et al., "Plate tectonics
     before 2.0 Ga: Evidence from paleomagnetism of cratons within supercon-
     tinent Nuna," *American Journal of Science* 314, no. 4 (2014): 878–94.

28.  Yebo Liu, Ross N. Mitchell, Zheng-Xiang Li, Uwe Kirscher, Sergei A. Pis-
     arevsky, and Chong Wang, "Archean geodynamics: Ephemeral supraconti-
     nents or long-lived supercratons," *Geology* 49, no. 7 (2021): 794–98.

29.  Richard Ernst and Wouter Bleeker, "Large igneous provinces (LIPs), giant
     dyke swarms, and mantle plumes: Significance for breakup events within
     Canada and adjacent regions from 2.5 Ga to the present," *Canadian Jour-
     nal of Earth Sciences* 47, no. 5 (2010): 695–739; Wouter Bleeker and Rich-
     ard Ernst, "Short-lived mantle generated magmatic events and their dyke
     swarms: The key unlocking Earth's paleogeographic record back to 2.6 Ga,"
     in *Dyke swarms—Time markers of crustal evolution*, ed. E. Hanski, S. Mer-
     tanen, T. Rämö, and J. Vuollo (London: Taylor and Francis/Balkema, 2006):
     3–26.

30.  Mitchell, Zhang, Salminen, et al., "The supercontinent cycle."

## CHAPTER FIVE

1.  Thomas R. Worsley, Damian Nance, and Judith B. Moody, "Global tectonics
    and eustasy for the past 2 billion years," *Marine Geology* 58, no. 3–4 (1984):
    373–400.

2.  Thomas R. Worsley, J. B. Moody, and R. D. Nance, "Proterozoic to recent
    tectonic tuning of biogeochemical cycles," *The carbon cycle and atmospheric
    $CO_2$: Natural variations Archean to present* 32 (1985): 561–72.

3.  Thomas R. Worsley, R. Damian Nance, and Judith B. Moody, "Tectonic
    cycles and the history of the Earth's biogeochemical and paleoceanographic
    record," *Paleoceanography* 1, no. 3 (1986): 233–63.

4.  Richard Damian Nance, Thomas R. Worsley, and Judith B. Moody, "The
    supercontinent cycle," *Scientific American* 259, no. 1 (1988): 72–79.

5.  See Christopher R. Scotese, "Atlas of future plate tectonic reconstructions:

Modern world to Pangea Proxima (+250 Ma), PALEOMAP Project" (2018): 1–35; http://www.scotese.com.

6.  Fabio Crameri, Valentina Magni, Mathew Domeier, et al., "A transdisciplinary and community-driven database to unravel subduction zone initiation," *Nature Communications* 11, no. 1 (2020): 1–14.

7.  João C. Duarte, Filipe M. Rosas, Pedro Terrinha, et al., "Are subduction zones invading the Atlantic? Evidence from the southwest Iberia margin," *Geology* 41, no. 8 (2013): 839–42.

8.  João C. Duarte, Wouter P. Schellart, and Filipe M. Rosas, "The future of Earth's oceans: Consequences of subduction initiation in the Atlantic and implications for supercontinent formation," *Geological Magazine* 155, no. 1 (2018): 45–58.

9.  Harm J. A. Van Avendonk, Joshua K. Davis, Jennifer L. Harding, and Lawrence A. Lawver, "Decrease in oceanic crustal thickness since the breakup of Pangaea," *Nature Geoscience* 10, no. 1 (2017): 58–61.

10.  S. A. P. L. Cloetingh, M. J. R. Wortel, and N. J. Vlaar, "Passive margin evolution, initiation of subduction and the Wilson cycle," *Tectonophysics* 109, no. 1–2 (1984): 147–63.

11.  Cloetingh, Wortel, and Vlaar, "Passive margin evolution."

12.  Steve Mueller and Roger J. Phillips, "On the initiation of subduction," *Journal of Geophysical Research: Solid Earth* 96, no. B1 (1991): 651–65.

13.  Jikai Ding, Shihong Zhang, Hanqing Zhao, et al., "A combined geochronological and paleomagnetic study on ~1220 Ma mafic dikes in the North China Craton and the implications for the breakup of Nuna and assembly of Rodinia," *American Journal of Science* 320, no. 2 (2020): 125–49.

14.  Christopher J. H. Hartnady, "On Supercontinents and Geotectonic Megacycles," Precambrian Research Unit, Department of Geology, University of Cape Town, 1991; Chris J. H. Hartnady, "About turn for supercontinents," *Nature* 352, no. 6335 (1991): 476–78.

15.  James Brendan Murphy and R. Damian Nance, "Do supercontinents introvert or extrovert?: Sm-Nd isotope evidence," *Geology* 31, no. 10 (2003): 873–76.

16.  Erin L. Martin, C. J. Spencer, W. J. Collins, R. J. Thomas, P. H. Macey, and N. M. W. Roberts, "The core of Rodinia formed by the juxtaposition of opposed retreating and advancing accretionary orogens," *Earth-Science Reviews* 211 (2020): 103413.

17.  Tanya Atwater and Joann Stock, "Pacific–North America plate tectonics of the Neogene southwestern United States: An update," *International Geology Review* 40, no. 5 (1998): 375–402.

18.  Christopher J. Spencer, J. B. Murphy, C. W. Hoiland, S. T. Johnston, R. N. Mitchell, and W. J. Collins, "Evidence for whole mantle convection driv-

ing Cordilleran tectonics," *Geophysical Research Letters* 46, no. 8 (2019): 4239–48.

19. Martin, Spencer, Collins, Thomas, Macey, and Roberts, "The core of Rodinia."

20. Paul G. Silver and Mark D. Behn, "Intermittent plate tectonics?," *Science* 319, no. 5859 (2008): 85–88; Z. X. Li, R. N. Mitchell, C. J. Spencer, et al., "Decoding Earth's rhythms: Modulation of supercontinent cycles by longer superocean episodes," *Precambrian Research* 323 (2019): 1–5.

21. Ross N. Mitchell, Taylor M. Kilian, and David A. D. Evans, "Supercontinent cycles and the calculation of absolute palaeolongitude in deep time," *Nature* 482, no. 7384 (2012): 208–11.

22. David A. D. Evans, "True polar wander and supercontinents," *Tectonophysics* 362, no. 1–4 (2003): 303–20.

23. Evans, "True polar wander and supercontinents."

24. Adam C. Maloof, Galen P. Halverson, Joseph L. Kirschvink, Daniel P. Schrag, Benjamin P. Weiss, and Paul F. Hoffman, "Combined paleomagnetic, isotopic, and stratigraphic evidence for true polar wander from the Neoproterozoic Akademikerbreen Group, Svalbard, Norway," *Geological Society of America Bulletin* 118, no. 9–10 (2006): 1099–1124.

25. Lauri J. Pesonen and H. Nevanlinna, "Late Precambrian Keweenawan asymmetric reversals," *Nature* 294, no. 5840 (1981): 436–39.

26. Nicholas L. Swanson-Hysell, Adam C. Maloof, Benjamin P. Weiss, and David A. D. Evans, "No asymmetry in geomagnetic reversals recorded by 1.1-billion-year-old Keweenawan basalts," *Nature Geoscience* 2, no. 10 (2009): 713–17.

27. Ross N. Mitchell, Taylor M. Kilian, and David A. D. Evans, "Supercontinent cycles and the calculation of absolute palaeolongitude in deep time," *Nature* 482, no. 7384 (2012): 208–11.

28. Hartnady, "On Supercontinents and Geotectonic Megacycles."

29. Chong Wang, Ross N. Mitchell, J. Brendan Murphy, Peng Peng, and Christopher J. Spencer, "The role of megacontinents in the supercontinent cycle," *Geology* 49, no. 4 (2021): 402–6.

30. Henry J. B. Dick, Jian Lin, and Hans Schouten, "An ultraslow-spreading class of ocean ridge," *Nature* 426, no. 6965 (2003): 405–12.

31. See https://speculativeevolution.fandom.com/wiki/Amasia.

32. See https://speculativeevolution.fandom.com/wiki/Amasia.

33. Anne Sieminski, Eric Debayle, and Jean-Jacques Lévêque, "Seismic evidence for deep low-velocity anomalies in the transition zone beneath West Antarctica," *Earth and Planetary Science Letters* 216, no. 4 (2003): 645–61.

34. Francis Nimmo and Robert T. Pappalardo, "Diapir-induced reorientation of Saturn's moon Enceladus," *Nature* 441, no. 7093 (2006): 614–16.

35. Helene Seroussi, Erik R. Ivins, Douglas A. Wiens, and Johannes Bondzio, "Influence of a West Antarctic mantle plume on ice sheet basal conditions," *Journal of Geophysical Research: Solid Earth* 122, no. 9 (2017): 7127–55.

36. Ted Nield, *Supercontinent: Ten billion years in the life of our planet* (London: Granta Books, 2008), 19.

37. A. C. Şengör, "Tethys and its implications," *Nature* 279, no. 14 (1979): 14.

38. Wang, Mitchell, Murphy, Peng, and Spencer, "Role of megacontinents in the supercontinent cycle."

39. Bo Wan, Fuyuan Wu, Ling Chen, et al., "Cyclical one-way continental rupture-drift in the Tethyan evolution: Subduction-driven plate tectonics," *Science China Earth Sciences* 62, no. 12 (2019): 2005–16.

40. Craig R. Martin, Oliver Jagoutz, Rajeev Upadhyay, et al., "Paleocene latitude of the Kohistan–Ladakh arc indicates multistage India-Eurasia collision," *Proceedings of the National Academy of Sciences* 117, no. 47 (2020): 29487–94.

### EPILOGUE

1. Robert M. Hazen, *The story of Earth: The first 4.5 billion years, from stardust to living planet* (New York: Viking, 2012).

2. Jafar Arkani-Hamed, "On the tectonics of Venus," *Physics of the Earth and Planetary Interiors* 76, no. 1–2 (1993): 75–96.

3. Shu Ting Liang, Lin Ting Liang, and Joseph M. Rosen, "COVID-19: A comparison to the 1918 influenza and how we can defeat it," *Postgraduate Medical Journal* 97, no. 1147 (2021): 273–74.

4. Paul J. Crutzen, "Albedo enhancement by stratospheric sulfur injections: A contribution to resolve a policy dilemma?," *Climatic Change* 77, no. 3–4 (2006): 211; Philip J. Rasch, Simone Tilmes, Richard P. Turco, et al., "An overview of geoengineering of climate using stratospheric sulphate aerosols," *Philosophical Transactions of the Royal Society A: Mathematical, Physical and Engineering Sciences* 366, no. 1882 (2008): 4007–37.

5. Francis Alexander Macdonald and Robin Wordsworth, "Initiation of Snowball Earth with volcanic sulfur aerosol emissions," *Geophysical Research Letters* 44, no. 4 (2017): 1938–46.

6. See https://www.abc.net.au/news/2017-11-18/plant-respiration-co2-findings-anu-canberra/9163858.

7. Chris Huntingford, Owen K. Atkin, Alberto Martinez-De La Torre, et al., "Implications of improved representations of plant respiration in a changing climate," *Nature Communications* 8, no. 1 (2017): 1–11.

8. Steven Chu, "Carbon capture and sequestration," *Science* 325, no. 5948 (2009): 1599.

9. Ning Zeng, "Carbon sequestration via wood burial," *Carbon Balance and Management* 3, no. 1 (2008): 1–12.

10. Jason T. Weir and Dolph Schluter, "The latitudinal gradient in recent speciation and extinction rates of birds and mammals," *Science* 315, no. 5818 (2007): 1574–76.

11. Mário V. Caputo and John C. Crowell, "Migration of glacial centers across Gondwana during Paleozoic Era," *Geological Society of America Bulletin* 96, no. 8 (1985): 1020–36.

12. Alfred G. Fischer, "Latitudinal variations in organic diversity," *Evolution* 14, no. 1 (1960): 64–81.

13. Weir and Schluter, "Latitudinal gradient in recent speciation."

14. J. Taylor Perron, Jerry X. Mitrovica, Michael Manga, Isamu Matsuyama, and Mark A. Richards, "Evidence for an ancient martian ocean in the topography of deformed shorelines," *Nature* 447, no. 7146 (2007): 840–43.

# INDEX

Page numbers in italics refer to illustrations.